EINSTEIN'S RELATIVITY

IT IS NOT WHAT WE ARE TAUGHT AND USE TODAY,
BUT IT IS MORE THAN WE RECOGNIZE

LUTHER L. NAYHM, PhD

Tarkas Press

Columbus

Published by Tarkas Press

Copyright © 2019 Luther L. Nayhm

All rights reserved. No part of this work may be reproduced in any form without the written permission of the author.

ISBN: 9781731529060

All great truths begin as blasphemies

— George Bernard Shaw

The first principle is that you must not fool yourself — and you are the easiest person to fool.

— Richard Feynman

A great deal of my work is just playing with equations and seeing what they give.

— Paul A. M. Dirac

Contents

INTRODUCTION 1
CHAPTER 1—STURM UND DRANG 11
CHAPTER 2—BACKGROUND TO THE ORIGINAL KINEMATICAL THEORY 31
CHAPTER 3—THE ORIGINAL KINEMATICAL THEORY 45
CHAPTER 4—THE ORIGINAL ELECTROMAGNETICS THEORY 55
CHAPTER 5—THE DOPPLER AND RADIATION PRESSURE 69
CHAPTER 6—THE DOPPLER PLUS THE ABERRATION 91
CHAPTER 7—THE GENERAL THEORY OF RELATIVITY 109
CHAPTER 8—EINSTEIN'S RELATIVITY: SOME CONCLUSIONS 119

APPENDICES 127
APPENDIX 1—DYNAMIC RADIATION PRESSURE MODELS 129
APPENDIX 2—NOTES ON THE DOPPLER AND ABERRATION MODELS 161
APPENDIX 3—COMPTON'S SCATTERING MODEL 173
APPENDIX 4—PLANCK'S DYNAMIC RADIATION LAW 181

REFERENCES 187
REFERENCES 189

Introduction

This book is not an attempt at bashing Einstein or his physics or an attempt at describing Einstein the person. That Einstein was brilliant and innovative cannot and will not be denied, but he did make serious mistakes in his role as a physicist that have not been properly addressed. These mistakes sent modern physics on an errant course from which it has not ether deviated since Einstein's active years during the early 20th century or shown any signs of recovery.

In the spirit of full disclosure, I was never a fan of relativity, any part of it. I viewed it as a mistake. Fortunately, I never had to work with it...except that I did! My professional contact with relativity was with the Doppler expression, which came from the 1905 paper by Einstein, "On the Electrodynamics of Moving Bodies." I was intrigued by the question of how that single paper by Einstein had given rise to one of the most useful practical models, the Doppler, while simultaneously giving rise to what I consider as the greatest intellectual debacle in modern physics.

Consequently, I decided to look critically into special relativity. When I did, I was surprised by what I discovered, which was that there was no modern and detailed research that specifically addressed Einstein's 1905 paper in toto. It turns out that not going back to seminal works is a mandate within physics and such re-vetting of these early works is a type of social proscription within academic physics.

I go into considerable details on how and why physicists never go back and why there is no modern re-vetting of Einstein's seminal 1905 paper on relativity. It turns out that what is written about special relativity in our text books and elsewhere is based on a different formulation of special relativity produced by Hermann Minkowski, who introduced more modern mathematics into relativity in ~ 1907 than were employed by Einstein. These more modern mathematics changed relativity and not for the better. Nobel Laureate Steven Weinberg might identify Minkowski's reformulation as post-Einstein relativity, though it occurred without a proper gestation period to

Introduction

fully understand what relativity actually was. Post-Newtonian physics, for example, required about two-hundred years to emerge in its modern "Newtonian" form.

Einstein's 1905 paper was rejected during its own time, and what we think Einstein created was not his but was Hermann Minkowski's reformulation of the 1905 paper. Such reformulations have occurred many times as our mathematics and knowledge evolve in many areas of physics. The consequence of Minkowski's success, on the other hand, was to essentially focus all attention on the kinematical theory at the expense of the electromagnetic theory. In dissecting the 1905 paper, we keep the two fundamental assumptions of relativity, but on top of these were other assumptions that propelled Einstein forward as he wrote the 1905 paper. It is these assumptions that gave form to relativity and which can be successfully challenged. It is within these challenges that the kinematics of relativity is identified as misrepresentations of electromagnetic physics acting on charged particles.

At this point, I will slip into a sidebar and build on the sentiments expressed by Gordin in his book *The Pseudoscience War* and supported by many others regarding what I call scientific truth. Considerable intellectual capital is expended on the philosophical ruminations over pseudo, fake, or fringe science. As often stated about pornography, people claim to know it when the see it even if they cannot articulate what the demarcation point may be between pornography and something that is simply lascivious. So it is with science. As described by Collins, et al., the consensus on which science or author fits these categories that are defining illegitimate science is often driven by so-called paradigms, a slippery term coined by and variously defined by Thomas Kuhn. There are many areas of science that are at first debunked before evidence accrues to show that the suggested rejections of ways forward are, in fact, legitimate.

From a more practical perspective, new ideas are often rejected for more menial and venal reasons. Some of these reasons are the classical not-invented-here psychology, or the more common reason is that most people are unimaginative and do not tolerate ideas that undermine their own concepts of truth, especially if their social and professional status is built upon ideas or concepts being attacked by someone. Take, for instance, the quotes on the front plate of this book. Start with Shaw's quote: "All great truths begin as blasphemies." This could be a poster statement of fringe thinking. Next, the quote by Nobel Laureate Richard Feynman: "The first principle is that you must not fool yourself — and you are the easiest person to fool." This would describe how the consensus perceives one of the blasphemies challenging some particular line of thought. Though another quote by Feynman is more telling: "I think I can safely say that nobody understands quantum mechanics." This is a very telling quote from a man who was awarded a Nobel Prize for research whose foundations are quantum mechanics. But, the final quote by Dirac, another Nobel Prize winner and the father of relativistic quantum mechanics, is even more revealing: "A great deal of my work is just playing with equations and seeing what they give." This quote forms a demarcation within physics, in which the line between fringe science and mainstream science becomes blurred.

Many theoreticians within physics have been promoting the perspective that observational evidence should not be required for developing a valid theory. Einstein's held this belief with regards to general relativity, and his influence can be felt over the hundred years since the development of general relativity. Some physicist, such as Frank and Gleiser, among others, are writing about this phenomenon with some trepidation, since physics is fundamentally an observational science. At what point can we identify a demarcation between mainstream and fringe science when all that these theoreticians are doing is playing around with certain sets of equations?

In writing this book and *Newton's Gravity*, my biggest concern was that my efforts would become labeled as pseudo or fringe physics. Consequently, my research was restricted to investigating the who, what, where, when, why, and how of various elements of physics, first with regard to Newton' gravity and now with respect to Einstein's relativity. The results are consistent with how history of science academics work, with an added component of actually understanding and critiquing the science being researched to uncover omissions and missed interpretations and the consequences of these particular errors. I am not inventing new science. I am describing and clarifying old science.

However, I have further researched supporting physics, some of which is summarized and reported by academics who do fall into the blurry demarcation between mainstream and fringe science. Their work is useful because these academics are observers and have taken note of documented discrepancies between what has been measured and what has been baked into certain mainstream science. The conclusions of these academics can be categorized as borderline fringe, though that is just their interpretations of what has been observed. The point is that the observations are typically mainstream but often ignored. We have, consequently, developed an unhealthy attitude toward those who criticize relativity, partly because many of the critiques are little more than rants. And, to be charitable, it is often clear that the critics' reach has exceeded their grasps.

As it turns out, using the hindsight of a hundred years, there are many legitimate observations that can be used to upend the blind acceptance of relativity. These will be noted, but the objective of this book is not to develop a whole new form of relativity but only to show that the currently accepted form of relativity simply does not follow from the observational evidence available during the fin-de-siècle period. In addition, there were other forms of relativity, one called Lorentz Relativity, none of which will be vetted. Lorentz apparently ceded the invention of relativity to Einstein, but certain modern critics of Einstein's version make the claim that perhaps Lorentz Relativity should have been the early choice. I have no intention of following this line of speculation. A key tenant of Lorentz Relativity was that the aether was stationary in the presence of mass, which was supported by the fact that Michelson and Morley did not detect the motion of the Earth in their interferometer. This particular result is described in more detail in Chapter 2.

It is also true that certain arguments are refuted by using what I might call flawed science. One particular instance is van Flander's claim that the speed of propagation of the gravitational force is 10^6 c, where c is the speed of light. Such a high speed has been required for Newtonian gravity but not for general relativity, where Einstein identified the speed of propagation of any gravitational effects as being c. The arguments against van Flander's assertions and analyses require the use of special and general relativity, but if those two theories are flawed, we must re-evaluate van Flander's conjecture and analyses, which can a priori be labeled as fringe science. In fact, I find van Flander's efforts to be almost illiterate. On the other hand, Brillouin, who I referred to in more detail in Chapter 1, has commented that the choice by Einstein of the speed of propagation of gravity as c was very arbitrary. I don't know, and I do not intend to address such conundrums. My approach is restricted to re-evaluating the 1905 paper by Einstein and discussing what we can say about the 1905 paper and Minkowski's reformulation of the physics within the 1905 paper.

A comparison of the 1905 paper with Einstein's later interpretation of that paper can be found in Einstein's book, *Relativity: The Special and General Theory*, which is a public domain (in the United States) downloadable Pdf file. This 1916 book has adopted Minkowski's reformulation of relativity rather than Einstein's own formulation from the 1905 paper. However, in this present book, we present a classical interpretation of the Kinematic Part of the 1905 paper and the evolution of that paper resulting from Minkowski's reformulation, which ultimately led to the general theory of relativity. The Electromagnetic Part of the 1905 paper is not discussed in any detail in Minkowski's reformulation, and we will show that this is a significant blunder in modern physics.

The physics described by Newton set in motion a two-century-long digestion process that gave form to what we know as Newtonian physics, and by the fin-de-siècle period, observations were being made that could not be explained by these evolved Newtonian physics. Unfortunately, the new Maxwellian electrodynamics had not had the luxury of a similarly long digestion and evolution into the modern post-Maxwellian physics that we understand and use today.

In this book, the notion will be examined that the immaturity of Maxwellian electrodynamics, in part, led the new breed of fin-de-siècle physicists to focus their attention on both the mathematics and the kinematics, topics they understood well, at the exclusion of taking electrodynamics as seriously as they should have. The early and restricted view of what the new physics should be resulted in mistakes and omissions that have steered modern physics down many wrong paths, where modern does not just mean post-Einstein physics, it means modern in the sense of how we do physics now and how the culture in academic physics has become dysfunctional in lockstep with an emerging dysfunction in modern physics itself.

It is within Einstein's 1905 paper that we find the true but unrecognized core of relativity. Interestingly, it is straight forward to dissect Einstein's relativity to lay bare this true relativity. However, one cannot get to the true underpinnings via Minkowski's mathematics, but the critical underpinnings are accessible in the 1905 paper, before they were obscured and shielded by Minkowski. We need little more than algebra and a willingness to challenge the underpinnings to what have proven to be the false observations, false ideas, and fabrications that prompted Einstein to develop or, rather, to advance special relativity in the first place. In most criticisms of relativity that I have read, no one challenges the various driving beliefs, and in not challenging these beliefs, we are swept up in seemingly incontrovertible mathematics, mathematics that is so consistent that the critics must resort to their own fabrications and misinterpretations to challenge what they are criticizing.

Introduction

In this book, we discuss how to challenge the putative special relativistic assumptions to show how the kinematical theory is not only a fabrication, it is a misinterpretation of the electrodynamics that Einstein addressed in the second part of the 1905 paper and omitted from his 1916 book. The issue is one of poor-quality observations and subsequent misinterpretations of those observations. The quality of observations will be discussed in considerable detail in subsequent chapters. But we can note here that there were two key observations of physical processes that motivated the search for answers outside of classical Newtonian physics. One was the inability to detect the aether, which we discuss in Chapter 2 in considerable detail, and the second was the apparent increase in mass of electrons as they were accelerated, which is discussed in Chapter 4.

Addressing the second observation first, anything that is accelerated has been subjected to a force, and for particles the applications of forces start with electromagnetics, either as classical waves or as photons. In 1900, the only charged particle known and in use was the electron. The electron could be accelerated with an electric field or by an electric potential. The scientists of the time already knew how to relate the voltage and electric field to the acceleration of the electron, but the ideas of an electric field and potential were still being worked out and, as mentioned before, had not gestated long enough. When an electron is accelerated by an electric field, our modern understanding and Einstein's understanding as stated in his 1905 paper was that the driving electric field clearly experience a Doppler effect that can affect how an electron is accelerated. But, at that early time, neither the idea of electrodynamic fields nor the electrodynamic Doppler were well-developed concepts.

The mass gain of high-speed electrons is often associated with Einstein's mass-energy relationship, only Einstein never invented that relationship. What Einstein showed was that, using his equations, the kinetic energy of an object can be related to its mass, and an increase in kinetic energy means an increase in mass. The well-known relationship $E=mc^2$ was actually a constant of integration for a kinetic energy equation and was necessary to account for the kinetic energy of an object as the speed of the object approached zero, where there was no kinetic energy. We will show where this mass-energy relationship actually comes from.

It is one of the working hypotheses in this book that the immaturity of electromagnetic theory was a root cause for early misinterpretations that electrons gained mass as they were accelerated, which occurred because the electrons were not being accelerated by as much as predicted by the classical electrical force on the electron. In fact, we will show that it is the Doppler shift of the accelerating field that gives the mistaken appearance that the electrons gained mass the higher the velocity of the electron.

As to the first observation, at that time we were not able to detect the aether, which was a consequence of the Michelson and Morley experiment not detecting any apparent motion of the Earth. Consequently, Lorentz and Fitzgerald independently developed the hypothesis that length contracts in the direction of motion. A consequence of the Lorentz contraction was time dilation. Time dilation was a consequence of how time was measured using light pulses traveling certain paths. Time dilation in special relativity states that a moving object experiences a slower passage of time than would a stationary object. This led, among other things, to the discussions of reference frames and of which object was actually moving relative to another object?

The equations for time dilation…the slower passage of time in a moving frame, where frame means a reference coordinate system to which the characteristics of an object are referenced…are symmetrical, so that each clock in a frame is measuring a passage of time that is slower than in the other frame. Such nonsense leads to the idea that two clocks moving relative to one another both experience the same slowdown in the passage of time. Two spaceships far from any reference frame but passing each other and each referencing their speed to the other would experience the same time dilation. This paradox can only be sorted out with a master reference frame to which each space ship's speed is referenced. The conundrum of which clock is actually running slower led to what is called the twin paradox. The twin paradox can only be resolved by resorting to hand-waving arguments about which space ship is accelerated. We show how this paradox is actually resolved.

Einstein was brilliant, cantankerous, and argumentative, but he was also an innovative thinker. He was willing to take risks and the historical record shows that he redid his work to fix errors or flaws that were pointed out. He single-handedly introduced tensors…via his friend and mathematical mentor, Marcel Grossmann…into physics. But he ultimately fell victim to his fondness for an esoteric philosophy that eulogizes the beauty of the mathematics. To Einstein, the mathematics was simply an opportunity to manipulate mathematical models, and he had little or no a priori concern what the manipulations might represent or even if they represented

Introduction

anything, though he often had a goal in mind and continued to manipulate until some esthetically-pleasing or useful results emerged.

The quote by Dirac at the front of the book supports the thesis that many if not most of the theoreticians "played around" with equations to see what would emerge, and this approach is evident in Einstein's formulation of general relativity. He had no reason to create that mathematics, since there was no observational evidence pushing him toward general relativity. Einstein was motivated by some mathematical issues to find another way to explain gravitational attractions with representations that met certain requirements of mathematical expressions. By about 1919 he essentially abandoned any further pursuits in general relativity unless dragged back into the discussions by some sycophant trying to ride on his coat tails.

The negative results of the Michelson-Morley experiment and the subsequent concepts of length contraction provided a pathway for the development of what is called the equivalence principle. This principle states that it is impossible to distinguish between acceleration resulting from gravity and acceleration resulting from mechanical means, such as rocket acceleration. This equivalence allowed Einstein to develop an equation of motion based on the curvature or metric of the four-space, as defined by Minkowski in his reformulation of relativity, in which the motion occurs within a gravitational field. If the mass is removed or made small, the equations reduce to Newtonian motion. Since a falling motion in a gravitational field is, because of equivalence, the same as inertial motion in empty space, Einstein was able to introduce inertial relativistic ideas into a non-inertial system.

The curves identified above are called geodesics, which are the shortest paths between two points, which also turn out to be the least energetic paths possible between the two points. Here energetic means the work need to move an object between the two points. The medium, the geometry, or the presence of a field play the role of defining the geometric description of the path that photons or physical objects take in moving between any two points. What we will show is that there was a misinterpretation of the putative special relativistic effects such as spatial contraction, which renders the equivalence principle invalid, in addition to which we show that the four-vector defined by Minkowski is a fabrication, rendering the elegant work by Einstein in developing his general relativistic theory a wasted effort.

However, this book is not about general relativity but rather is about the special or restricted relativity. Without special relativity, there is no general relativity. The irony is that in the modern era there is no special or restricted relativity, only the general relativity that subsumes special relativity. It is a case of the child begetting the parent and then being abandoned by that parent.

Part of the issues with respect to criticizing relativity has been that there is nothing, so the narrative goes, to replace its apparent kinematical successes. However, contrary to that belief, there is something with which to replace it, and that replacement ironically comes from the electromagnetics of relativity. The research for this book revealed that the true special relativity was the electrodynamics that Einstein described in the 1905 paper. It is within the true relativity that we also identify new physics.

What we show is that the true but hidden relativity forces us to re-evaluate exactly what our experiments and observations are revealing to us. Plus, we show…or at least make the case for…re-evaluating the technologies we use for making observations purporting to support relativity and high-energy physics. In addition, we identify new technologies and re-interpretations of existing technologies and discuss the limits of these technologies. Whether the discussion on the kinematics is persuasive does not invalidate the discussion on the electromagnetics, which is surrounded by misinterpretations and omissions, many of which are discussed in Chapters 4 and 5 and in Appendix 1.

Consequently, we will supply a description of the current beliefs in relativity before diving into the devilish details that should put relativity as we currently understand it to bed once and for all. Because of the contradictions both within relativity and with the criticisms of relativity, the first part of Chapter 1 is both historical and verbose. Nothing presented in this book are unknown but what is known is subjected to a re-evaluation considering a centuries' worth of digestion and a willingness to perform the re-evaluation.

Chapter 1—Sturm und Drang

The key to understanding what relativity really is requires parsing relativity into three distinct theories, two of which are wrong and one of which is more than we have understood. There is a natural progression in discussing these three theories, beginning with a thorough discussion of the first and so-called Kinematic Part of the 1905 paper and proceeding to the second or Electromagnetic Part, which turns out to be the only real part of relativity. The third discussion is a summary of general relativity, which is brief because it will become clear from the first two discussions that there is no general relativity, at least as currently understood.

As mentioned previously, there has not been a direct modern review and critique of the 1905 paper, only of Minkowski's reformulation. And as also mentioned, even if the arguments and critiques of the Kinematic Part of the paper are not persuasive, the examination and discussion of the Electrodynamic Part of the 1905 paper can essentially stand on their own as a revelatory critique of missed and misinterpreted physics with exciting potential in remaking our understanding of certain technologies and physics.

To begin, Chapter 1 is basically a review of the origins and critiques of special relativity as we currently understand it. If the reader wants more details, they will need to dive into some of the references to understand the supporting physics that was the underpinning to the creation of the theory of relativity in the first place. The old arguments both pro and con about special relativity are made by people who simply do not understand the physics and, in some cases, do not want to, since as with other critics of modern physics, they are true believers and have no intention of or capabilities to expose the fact that relativity as we understand it is incorrect on the one hand and incomplete on the other.

Chapters 1 and 2 were especially interesting to produce, since, until I began to pry into the history of the development of special relativity, I was only vaguely aware of the degree of both adoration and antipathy that were directed toward both Einstein and his relativity. There are many balanced discussions of the various roles played by many of the luminaries in physics during the fin-de-siècle period during which relativity, photon theory, and quantum mechanics were emerging. But this was one hundred years ago, and we have many more insights into the issues with which the physicists of the time were struggling. It is easy to criticize but it is harder to create, and my role is simply as a critic using the hindsight of a hundred years of physics and technology to support my research.

In describing various elements of relativity, we will occasionally use the phrase "not even wrong". The origin of this phrase was the acerbic Nobel Laureate Wolfgang Pauli, who lavished it as an epithet upon the unwary who approached him with ideas so stupid to his way of thinking that one could never prove or disprove those ideas. This phrase was picked up by Peter Woit in the title to his book criticizing modern field theories. Others have also used it in conjunction with criticisms of what they call pseudo-science, one of whom is Michael Gordin in his nifty book *The Pseudo-Science Wars*. Gordin's point of view emerges elsewhere in this book.

There have been many thousand papers published on special relativity, and they are still being published. Relativity is far from a settled matter. From my perspective, something must be missing to cause such confusion, contradiction, and puzzlement after over a hundred years. I read dozens of papers and many advanced texts in preparation for writing Chapter 1, and the documentation I found most interesting were those associated with the history of the origins of special relativity. Everything else can be found in a variety of current textbooks and in many well-written "classics" available at reasonable prices from Dover Press. Arguably, special relativity is frozen in its ~1908 form and general relativity is frozen at most at its ~1916 form.

Of the books written about relativity, the one written in 1921 by the 21-year-old Wolfgang Pauli may be one of the best. It was finally translated from German to English in the late 1950s and is available from Dover Press. This was likely the first of such books that were written in the early 20th century and indicates what the prevailing wisdom concerning relativity was and has been for nearly a hundred years. The book is very readable as such books go but should be compared with the 1905 paper by Einstein, also published in a Dover Book among other sources. It is interesting to compare the clarity of both Pauli's and Einstein's works as well as how the content and focus in discussing relativity was changed by Minkowski away from the presentation within Einstein's 1905 paper. Pauli's book reflects Minkowski's reformulation.

Unfortunately, most refereed papers, reviews, histories, and texts relating to relativity have been written by true believers, believers in Einstein's infallibility and believers in both his special and general relativistic papers, though there are also those who are skeptical of the idea that Einstein was an intellectual superstar. Many of the critiques have been written by skeptics who present their views outside of any refereed sources, often from lecture or symposia notes with only a few exceptions, but most such critics have not put forth substantive arguments to support their positions. Ironically, many of the fringe researchers also believe in relativity but use its complexity to support their own views of related physics, which is where they demonstrate their unorthodox research.

My perspective is that of a skeptic, but until I unraveled the putative physics for myself, I simply did not care one way or the other about relativity. For the true believers, the truthiness of special relativity is self-evident. As with Newton and his gravity law, I went far enough back into the literature to identify what mistake, misinterpretations, or omissions may have occurred.

Steven Weinberg has made the case in his review of Kuhn's 1962 book, *The Structure of Scientific Revolutions*, that scientist, especially physicists, are subjected to certain paradigms that emerge as any give area of science matures into its accepted modern form. This may be as much a cultural imperative as it is a consequence of natural intellectual activity. To paraphrase, we never go back to the seminal works because they are too difficult to comprehend, so we do not do it. As shown in *Newton's Gravity*, we accept paradigms at great peril to the progress and verisimilitude of subsequent science.

However, to carry Weinberg's philosophical ruminations a bit farther, it is useful to interject my own experiences into the discussion. As a student, I was told what relativity was and it was Minkowski's formulation, though it was not identified as such. It was not until much later that the original 1905 paper was available to me and even then, I was not particularly interested in it, partly because it is poorly written and partly because I had already come to the opinion that relativity and much of the other advanced physics were a misinterpretation of observations. My professional experience led me to that latter conclusion. However, in retrospect, I should have been aware of that weakness in modern physics earlier in my career.

The question may arise as to why the research I will be describing in this book has not already occurred. Why have physicists not gone back and performed the research that I am presenting? Basically, they cannot. Academic physicists are coached that certain topics, such as those in Einstein's 1905 paper, are done-deals and any critiques of that paper is a career-ending event. By the time physicists have either achieved tenure or chosen some other career path, there is no longer any academic interest in relativity. On the other hand, many tenured physicists at the twilight of the careers begin to leave the mainstream in their research, with many drifting into what is labeled as fringe science. And non-academic physicist, who have extra-academic careers, are simply trying to apply what they have been taught…there is no benefit or payoff in critiquing what has been taught. In fact, according to Weinberg, there is no reason for any such critiquing…there is nothing there to find.

What is generally not appreciated about academic physicists is how arrogant, insular, and snobbish they are as a class. The book by Ann Finkbeiner, *The JASONS: The Secret History of Science's Postwar Elite*, contains an excellent exposé of academic physics, though that was not what she intended. Such luminaries as Stephen Weinberg are reinforcing a sentiment silently imposed on graduate students as if by a hidden hand. We are coached that there is nothing in the fundamental papers that has not already been recast in more modern form, and is, consequently, more accessible than in the original papers. We are coached that there is no career path in delving into the earliest fundamental efforts. We are coached to argue…sometimes viciously…about what we know when that knowledge is challenged. And we are coached not to challenge those who have by agreement of the consensus achieved some level of achievement. Nobel Laureates are seldom challenged or second guessed. Unlike in almost any other scientific discipline, academic physicists never challenge the foundations to their disciplines. Those who are interested in history of science do not question the actual work they are researching, and, in the case of Einstein, the historians are all true believers. I was never a true believer.

This chapter and subsequent chapters have been synthesized from two different perspectives. One perspective is to present essentially verbal descriptions, text, and dialogs. The second perspective of the synthesis has been placed in an associated appendix that contains some mathematics that support the descriptions in the main chapter. The reason for this is that an understanding of what will be simply stated as facts in the main chapters requires an analytic basis for being understood. The mathematics support the dialog, which otherwise would be indistinguishable from bloviating. Most non-technical people simply have lost their ability to follow the mathematics as present even in our introductory texts, and most non-technical people are challenged even by their scholastic mathematics. No such knowledge, however, is required in reading such books as Walter Isaacson's book, *Einstein: His Life and Universe*.

Einstein's Relativity

In reading Isaacson's book, which I enjoyed, we see that Isaacson is simply a reporter. When he reports on what some physicist has spoken to him regarding relativity or Einstein, Isaacson is reporting the verbal descriptions just as a reporter would. He does not require any mathematical descriptions to justify the narrative and we take it as a matter of faith that he is being literal and factual. That is the level at which most people understand relativity, which is also the level presented in science fiction. I need to go further in this book, since I am not just reporting on relativity, I am challenging major parts of it as well as all Minkowski's reformulation. Still, it only requires undergraduate mathematics to challenge relativity as described in Einstein's 1905 paper. That being case, I will insert high-school algebra into the discussion, assuming...likely erroneously...that most people who read this book will know how to multiply, divide, and write and understand a simple equation, such as $v = a\,t$.

This first chapter is focused on the fundamental ideas and observations that motivated Einstein and his peers to construct the theory of relativity, a term applied in 1906 to the physics described in Einstein's seminal paper published in 1905, "On the Electrodynamics of Moving Bodies." However, regardless of the initial motivations for pursuing the theory of relativity, research into the origins of special relativity lead to a series of modern papers and research that have essentially been discounted or ignored by the bulk of physicists, if they are even interested in the topics. It is easy to independently arrive at similar conclusions as the referenced research, though the approach is different. Unfortunately, many of the modern papers are from fringe publications and from fringe researchers. None the less, it was not their work but their sources of observations that were most useful. What was discovered in reading some of these documents is that there are claims based on their own research that could be explained by overlooked aspects of Einstein's work on special relativity. We will identify these points later as they arise.

None the less, the modern research essentially undermines some of the experiments that are purported to have supported the initial development of relativity. I will proceed to describe some of the more modern work that is referenced and that I am referring to, if only to show that all roads lead to the same endpoint, which is that there is nothing special about special relativity, except in the overlooked electrodynamics. To quote Gertrude Stein, "There is no there, there." Or was there? As it turns out, there is some there, there but not where we think it is.

As this chapter unfolds, I will defer certain discussions to later chapters. The reason for this is that the kinematics in relativity will be seen to be a misinterpretation of the electrodynamics in the 1905 paper, and the electrodynamics are not addressed in detail until Chapter 4. In the current chapter, I will state some findings and conclusions reached in the later chapters, but I will not dwell on those physics until later.

As I was completing the discussions in this chapter, I came across a small book by Léon Brillouin, *Relativity Reexamined*. Brillouin was a prominent (and perhaps the greatest) and the preeminent French physicist within the first half of the 20th century and Poincaré's natural successor. Brillouin discusses and puzzles over many of the same issues within relativity as I have, though I came to different conclusions in many instances than did Brillouin. Our conclusions differ because I had more information available to me than he did at the time of his death in 1969 and because I dug deeper into the underpinnings of special relativity. None the less, I found that Brillouin ruminated over many of the same shortcomings of relativity as I had in complaining that certain things were stated as fact in the 1905 paper when they were no such thing. They were assumptions and statements made with no justification, but this gets ahead of the story.

Brillouin believed in relativity but not in many of the assumptions associated with it. He felt that the two theories, the restricted or special relativity and the general theories, were incomplete. In this book, I express the opinion that there is little correct about either the restricted or general theories as they are currently understood, and what is correct is ignored.

Brillouin appreciated that physics is fundamentally an observational discipline. Where he and I differ is in how we interpret what has been observed, which I will follow up later in this and subsequent chapters. None the less, a quote by him is revealing: "The laws of classical mechanics represent a mathematical idealization and should not be assumed to correspond to the real laws of nature. We now have to realize that errors are inevitable (...) a discovery that makes strict determinism impossible." This perspective was carried over into his critique of relativity and his assessment of quantum mechanics.

Brillouin discussed certain elements of relativity with regards to information, though not detailed in the referenced book of his that I read. A key observation in my own analyses is that Einstein used incomplete models to describe the physics that he was interested in. Brillouin also concluded that most of Einstein's models are first-order approximations, a conclusion with which I agree. Throughout this book, I harp over and over on the observation that Einstein does not develop his models with real measurements in mind. Einstein's models are, therefore, idealizations...at best. In addition, Einstein engages in an overreach in applying physics from one sub-discipline of physics in another sub-discipline. In particular, there are many ideas within electrodynamics that have their analogs within kinematics, and Einstein and Minkowski erred in convolving these analogs.

I go further than Brillouin and identify certain of the driving assumptions for special relativity as being bogus in that these assumptions are based on incomplete or incorrect information. These are not the two famous postulates relating to the speed of light being frame independent and to the proposition that all laws (representations) of physics are frame independent, ideas first put forward by Henrik Lorentz. And by frame we mean any coordinate system that may be moving linearly relative to any other coordinate system. While these postulates pre-date Einstein and are the two fundamental postulates, there are many more assumptions used in developing relativity.

Some of the postulates or assumptions are now questionable. The origin of these assumptions arose over many years and were part of the erroneous consensus that had been established by the time Einstein wrote the 1905 paper. One postulate was that a moving observer could not know they were moving if they were moving inertially...not accelerating...and not making observations external to their frame. This postulate arose from the so-called null result for the Michelson-Morley (M-M) attempt at measuring the movement of the Earth through a fictious medium called the aether. The M-M experiment is discussed more completely later. The interpretation of the results of the M-M experiment means that any observations or measurements made inside a black box cannot determine if the box is moving without acceleration. The resulting hypothesis or postulate was that, without cues coming from the outside, the observer could not know they were moving.

The postulate or assumption that we cannot know we are moving as defined above is fundamental to other ideas that were not necessarily part of special relativity, such as the notion that we cannot distinguish gravitational attraction from the force associated with being mechanically accelerated on a space craft. Another postulate is that we cannot distinguish free fall in a gravitation field from free fall associated with inertial coasting in a space craft. These were combined in what was known as the Equivalence Principle, which is fundamental to general relativity. What will be shown is that, if we can, in fact, measure that we are moving when we are in a closed box, the consequence is that these other two assumptions can be proven false. We will revisit the Equivalence Principle later in discussion related to general relativity.

Finally, there is a concept of simultaneity that has driven the evolution of relativity. If two separated events happen simultaneously in an inertial frame and as measured by someone in a different inertial frame (moving at fixed speed and not accelerating), then someone in the moving inertial frame will not measure these two events as being simultaneous under all but a single accidental circumstance. Therefore, time, so it is stated, is relative and not absolute. In more familiar terms, if two events on the ground are synchronized, such as sending out radar or radio pulses from two different but synchronized locations, a cruising aircraft moving at a constant speed would not measure these two events as occurring at the same time, unless the aircraft is on the perpendicular bisector of the line jointing the location of the two events, and even then, the aircraft would not know if the events were synchronized. The question is, is this issue significant and are the conclusions reasonable?

Let's follow up on the above conjecture. If we continue to generalize the location and speed of the aircraft relative to the two synchronized radars, we would be developing various well-known scenarios used in tactical missions in which an aircraft with an arbitrary location and speed is engaging in resolving two active targets on the ground separated by some distance. As we add information to the aircraft through various sensors suites, we begin to resolve certain ambiguities. If we do not know and cannot find the range to the radars, we are left with unresolved ambiguities. We lack the information to determine such things as whether the pulses were transmitted simultaneously. Time rates have little to do with the lack of information, and without enough information we can only make relative measurements. Simultaneity was used to justify the idea that clock rates were not the same in frames moving relative to one another, and, consequently, the interval between the two detected pulses would be different in the two frames, which is independent of the clock rates in the two frames.

This particular theme regarding the of lack of information will be shown to be the underpinning to other issues related to relativity and to the various thought experiments used to develop the theory and equations that have become familiar. The fin-de-siècle physicists did not have radars, they did not have aircraft, they did not have communications systems, and their concept of the Doppler for electromagnetic radiation was incorrect. They struggled with new concepts, and the confusing issue was that electromagnetic radiation traveled with a constant speed in all moving frames. They were grappling with ideas that would not be resolved for nearly forty years, but then, because of the prevailing paradigm, nobody questioned Minkowski's reinterpretation of Einstein's relativity.

Generalizing, we find that many if not most of the assumptions in the 1905 paper are based on a naïveté and misplaced trust in the results of experiments that were then unavoidable but that we now know how to handle for many of the scenarios that Einstein contrived in his thought experiments. In modern terms, some scenarios are inherently ambiguous because of a lack of information. Some elements of relativity are intended to dismiss these ambiguities by contriving physics to do away with the ambiguities, such as the concept of simultaneity. But, more egregiously, we have ignored the modern observations that would show that the observations used by the fin-de-siècle physicists were the results of bad observations or limitations in their equipment and their abilities to perform adequate observations, plus the poor-quality and forced interpretations of the observations.

I have leaned on Brillouin's review of relativity to show that Brillouin's analyses were on the right path, but he did not take things far enough, since he was still operating on the assumption that relativity was a fundamentally valid effort that had been botched by Einstein. In that regard, he is correct, but he still clung to the idea of relativity because he was not willing to dismiss Poincaré's efforts out of hand. This theme is reinforced in the following historical discussions.

As with Newtonian gravity, there are several fundamental errors in special relativity that emerge upon reading the seminal 1905 paper written by Einstein. And, as with Newton's magnum opus, his *Principia Mathematcia*, most people rely on other people's interpretation of that paper, and if they even bother to read it, they don't understand it well enough to be able to properly vet the ideas in it even if they were motivated to challenges those ideas. Brillouin made these same observations, which effectively portrays many adherents to relativity as sycophants and poseurs.

What Brillouin missed is that the 1905 paper is only about electromagnetics, but more specifically it is about optics, which is electromagnetic, which is the specific form of electromagnetic propagation what was well understood at the time. Despite this, the first part of the 1905 paper, which is the part most people have heard about, is framed as kinematics and not about light…only, it is all about light! All the kinematics is really an erroneous paraphrasing and modeling of what is in the second part of the paper, which is almost all about the propagation of light. This error is discussed further in Chapter 3. In this chapter, however, we focus on the collection of issues that were driving Einstein toward special relativity, though the details regarding some of these assumptions are reserved for later chapters.

Criticisms of special relativity has taken many forms, often arising from mathematical perspectives but also often resulting from the petty and dishonest ways in which one person and their works are attacked by others. The Wikipedia article "Criticisms of special relativity" is a representative snapshot of the range and types of criticisms that permeate discussions of relativity. Often the proponents of certain viewpoints rely on experimental evidence that can now be discredited, some of which is discussed later.

Interestingly, in the development of general relativity, Einstein was criticized for his dismissal of experimental or observational evidence as being irrelevant and a side issue. We will come back to this general theme many times in this book. However, no one ever appears to have critiqued the detailed contents of the 1905 paper after Minkowski reworked it. And, in fact, Minkowski did not bother to review the 1905 paper even as he reformulated it. It is as if no one read the paper and understood it and then compared it with Minkowski's reformulation, which is consistent with Weinberg's and Kuhn's beliefs on paradigms and how these paradigms prohibit revisiting the seminal works in some scientific area, which is where the criticism in this book begins.

Unlike Newton's *Principia*, which has been translated from the Latin to English by many different people and groups, the 1905 paper has only a single "approved" translation that is prominently part of the scientific literature in English, and that translation did not occur until 1923. The English version is almost a literal translation of what Einstein wrote in turn of the century German, a time when dozens of dialects were still present. The English translations of the 1905 paper made little attempt to "Anglicize" the grammar and there have been many debates over exactly what was meant by one statement or another in that paper, including finding and parsing drafts or galley-proof modifications to understand Einstein's intentions in using certain approaches to various models. Some of the early interpretations of the 1905 paper leave out or modify what Einstein wrote, though without fully understanding what they were reading, it is unclear why the authors felt they should insert themselves into the physics. The reader of the 1905 paper is best served by simply wading through the complete 1923 published translation by W. Perrett and G. B. Jeffery. Most of the translations into the prominent languages of the time of Einstein's relativity papers were made in the early 1920s. Unfortunately, few of these translations were re-translated into English, so the rough edges in the Perrett and Jeffery translation cannot be compared with how others interpreted the 1905 paper. The paucity of English translations is indicative of the time it often took for ideas published in one language to percolate into another language and also indicates how limited the audience for that 1905 paper really was.

However, in 1916 Einstein produced a small book, *Relativity: The Special and General Theory*, in which we find both a much clearer interpretation of the 1905 paper and the evolution of this work arising from Minkowski's reformulation, which ultimately led to the general theory of relativity. In the 1916 book, Einstein attempted to explain in detail what the content of the 1905 paper was all about but in cleaner prose. None the less, he explained nothing new and clarified little. The 1916 book is much easier to read than is the translation of the 1905 paper by Parrett and Jeffery, yet the book is not a translation per se of the 1905 paper, it is a paraphrasing by the author using Minkowski's reformulation, analyses, and assumptions. The clarity of that book may be because it had the benefit of a different translator and editor, since the 1916 book was first written in German.

One observation is that the Perrett and Jeffery translation of the 1905 paper directly into English is rough at best. Either the translators were not very good or else the 1905 paper was apparently written in a style somewhat unfamiliar to them, even though fin-de-siècle German was the lingua franca of science until the post-WW II era, at which point English became the new lingua franca of science. None the less, casual reading of the 1905 paper's translation would miss a propaganda style in which the author did not actual do what he said or even pretended to do what he said. Or, more egregiously, simply stated ideas, concepts and approaches as facts with no support or justification, which was also noted by Brillouin. The propaganda overtones will make more sense as this book proceeds. The references supply a selection of sources for the 1905 paper.

It is worthwhile noting that Brillouin commented that even in the 1920s Einstein's scientific writing style was opaque, which lends credence to the notion that Einstein was simply not able to pin many of his ideas down into precise language. Nobel Laurate Richard Feynman made a general observation that an inability to simplify some discussion usually implied that the topic was not all that well understood. Brillouin was not a fan of Einstein and Einstein's writing style, though he references Einstein's genius while criticizing the vagueness and contradictions of certain fundamental assumption that Einstein made. Brillouin's reading of Einstein's writings indicates that the lack of clarity present in the 1905 paper seemed to persist throughout Einstein's career, at least in his scientific writings, which would indicate how heavily edited Einstein's more popular writings were.

It is also important to emphasize here that nothing useful regarding the physics of special relativity can be gleaned from the following summary of the criticisms of special relativity. Rather, we will begin to see something of the workings of a culture which often pretends to have little dirty linen and for whom the public faces of disagreements are often described as little more than gentlemen's arguments over port and cigars or the subject of occasional snide comments. The reality is that the infighting can become vicious and personal. While such behavior is seen to publicly erupt from various areas of academe upon occasion, it is generally understood that academic physics has cultivated and promoted these belligerent attributes as a way of life. Physicists are taught not to just be critical of the science under assault, we are taught to win every disagreement, which occasionally leads to what amounts to school yard fights that are usually out of sight of public scrutiny.

Still, as a German and a Jew, Einstein attracted unnecessarily vicious attacks that missed the physics to strike the man. David Kaiser's short article supplies some insights into the role of national and academic politics in how scientific ideas are disseminated or destroyed. It is often like modern presidential politics in the United States. Glashow's essay on Newton demonstrates that these same forces were in operation during Newton's lifetime and that Newton was himself a vicious infighter.

To begin, as Olivier Darigol points out in "On the Genesis of the Theory of Relativity," Einstein was far from the first or the last to write on the topic of relativity. Einstein's 1905 paper had no references and gave no direct credit for the topic of relativity, a situation for which Einstein was criticized, just as Newton was criticized two centuries earlier. The 1905 paper was a re-statement of the work of others plus Einstein's own novel contributions. Einstein owed much to those whose work laid the foundations for relativity. And, even though I am critical of relativity, I will also praise those unique contributions within the 1905 paper that will be shown to be the actual relativity.

Within modern criticism of relativity, there has emerged considerable vitriol in certain quarters over the topic of to whom the credit should go for "inventing" relativity. These altercations have persisted for a hundred years at one level or another, occasionally dying out only to spring back as one person or another smolder over ancient wrongs. These squabbles erupt because the antagonists are true believers in the efficacy of special relativity and the French are still smarting over the fact that Poincaré is simply noted as a major participant but not the discoverer of relativity.

Einstein apparently ramrodded himself into the discussion on relativity, and prior to publishing his 1905 paper had never produced any research or commentary on that topic. He and his paper were intensely disliked across a broad swath of fin-de-siècle theoretical physics. Darigol purports to supply an historical reading of the state of affairs during those times, but many of the links to Internet documents are no longer active. In fact, even the links within referenced journals and available documents are often broken. This is, of course, the risk of relying on the Internet for finding references, and often the reference to the original document is the best we can do.

My review of the literature relating to relativity showed that this literature contains a rat's nest of criticism, mostly about Einstein rather than relativity itself. Support of relativity is tacitly assumed through its apparent broad acceptance within the physics community. Almost no one has mounted a modern criticism of the physics, though if one were to trace the antecedents to certain modern field theories, criticisms at those levels is indirectly a criticism of relativity, though that perspective is essentially ignored. What we show is that these advanced theories are not even wrong, because the form of relativity they are based on is not even wrong, which is Minkowski's reformulation of the physics laid out in the 1905 paper. We do not go into this aspect of the criticism until Chapter 3 after introducing the electromagnetics of special relativity.

Brillouin identified many sources of confusion that have led to misinterpretations within special relativity, though he did not supply an analysis that pinpoints how the kinematics is derivable from the Electromagnetic Part of the 1905 paper. Brillouin did not critique the 1905 paper per se and it is not clear whether he believed that the fundamental foundations of relativity held. It appears he felt that the broad concept of relativity held whereas the details and interpretations were flawed and need rework and that certain models were simply first-order estimates of the actual physics. Brillouin's book did not generate much interest when it was published, but his work did highlight the multitude of conflicting interpretations and lack of observables that have vexed theorists for a century.

None the less, regardless of the background of the various authors who have attempted to explain or critique relativity, they believe in relativity without truly understanding it. The vitriol and ignorance toward Einstein and relativity are occasionally alarming. Even in the articles regarding who should get credit for relativity, all the authors are believers in the theory, so the discussions are simply about the priority and not the physics.

One French author, Christian Marchal, like many other French scientists, complains that, as some other critics imply, Einstein may have plagiarized Poincaré, the preeminent French theoretician of the time, and "stolen" relativity from him with the complicity of David Hilbert, a dominate mathematician at Göttingen University, and of Max Planck, who was at Berlin University and a future Nobel Laureate. Göttingen dominated German and, therefore, European theoretical physics during the fin-de-siècle period.

Per Marchal, Hilbert and Planck would do everything in their powers to see that German dominance in physics would be maintained…at any cost. Marchal paints Hilbert as a jealous and unrelenting critic of Poincaré and ultimately an unrelenting champion for maintaining German scientific dominance. Marchal further points out…or perhaps whines…that Poincaré was not recognized as a physicist and could not get published in physics journals, so his work was unknown within the broader community of physicists. Yet Poincaré's work in theoretical physics was well established, so one can conclude that certain jealousies are at work as are and were national pride. These themes have percolated throughout European cultures for centuries.

Poincaré's contributions were mainly in mathematical rigor or, in the case of general relativity, the non-rigor of classical gravity, and his accomplishments during the fin-de-siècle period were part of a growing trend adopted by Einstein, which was to believe that the mathematics are the physics and not simply tools used to describe the physics. Logunov has written a useful summary of Poincaré's and other's contributions to relativity and concluded that, as Darigol points out, special relativity had many parents and that Einstein's had put the finishing touches on something that was nearly completed by Lorentz and Poincaré. Rothman puts these ideas into a readable and well-researched paper that sorts out the contributions of the major players in the invention of relativity. As good as Rothman's discussion is, though, he is a true believer, and while he criticizes physicists for not actually reading the 1905 paper while purporting to display their knowledge of its content, he does not seem to have critically understood it himself.

As we will see, everyone is a true believer and, consequently, everyone has missed the point that Einstein was the sole parent of the electrodynamics that are the true relativity, though Marchal would dispute even this point. Brillouin also missed this point, though he felt that many things were being misinterpreted, but the core idea was to him sound if only poorly developed. On the other hand, Poincaré was an expert in the mathematics of electromagnetic theory and was, therefore, very knowledgeable in certain areas of physics, which likely further fueled Marchal's perception that even the electrodynamics in the 1905 paper had been lifted from work by Poincaré. However, I find no evidence that Poincaré was more than just focused on the mathematical rigor of electrodynamics, though I did not exhaustively research Poincaré's oeuvre. Among those who have studied Poincaré's contributions, he was certainly recognized as a major contributor, along with Lorentz, to the theory of relativity as it became known.

As far as I can see, Marchal is alone in the amplitude of his defense of Poincaré's position in the history of relativity, though there have been rumblings over the century, mostly from French scientists, that there was something fishy about when Einstein happened to publish his 1905 paper and about the content and origin of that paper. I have no opinion about these stated criticisms other than most are not based on a good criticism of the actual paper itself. I let the physics stand on its own and in doing so I find the 1905 paper to be in part a paraphrasing, in part a fabrication, and in part brilliantly original. It is the brilliantly original part that turns out to be the only relativity.

A much more balanced approach to the discussion can be found in Roger Cerf's article "Dismissing renewed attempts to deny Einstein the discovery of special relativity". Unfortunately, while that article was available as a Pdf file when I started this book, it is now a broken link, and the article is only available as a journal article for which one must either pay a fee or have access to a science library. This is simply one more example of the dynamics of links to articles available on the Internet.

A brief survey of more professional criticisms can also be found in the Wikipedia article relating to the "Relativity Priority Dispute," and the broken link to Cerf's article is supplied among those references. There was also considerable discussion in the 1950-80-time over Einstein's general relativity and on competing theories. Relativity, especially general relativity, was an unimportant area of research up until the late 1950s until more modern observations began to apparently, or perhaps putatively, require Einstein's relativity to supply models to explain what was being observed. However, amid considerably rancorous discussions, Einstein's general relativity finally became the Gold Standard for such theories, because none of the competing theories supplied predictions and results that were any more accurate or observable than those supplied by Einstein's theory. What will be discussed later is that these competing theories started at the same place as Einstein's theory started and, hence, suffered from the same flaws as Einstein's general theory.

The general theory is only mentioned here because it contains the special or restricted theory of relativity being discussed in this chapter. There is no longer a standalone theory of special relativity…it is simply a piece of the more extensive general relativistic theory, since in the absence of any mass…empty or nearly empty space…general relativity reduces to Newtonian gravity and kinematics…at least as we currently understand Newton's point-mass approach to gravity. If special relativity can be refuted, then general relativity is automatically refuted.

The point is that the claim that Poincaré had the ideas for the advanced versions of relativity first belies the fact that there have been a multitude of co-contributors, each adding incrementally to the general understanding of relativity. The concepts were being discussed openly at the time. None the less, Darigol's essay supports the dismissive and ignorant attitude among many that is still present in terms of what Einstein really accomplished in 1905. As with Newtonian gravity, the concepts were being widely discussed among a relatively small clique within physics and many papers and experimental results were already being published and discussed prior to Einstein's 1905 paper, and many of these experiments presented unexplained results that special relativity addressed.

I decided to take the high road within this discussion. Consequently, rather than simply engage in a diatribe over special relativity and follow the uninformed historical criticism, my approach is to show that Einstein's physics is completely wrong on the one hand and creatively brilliant on the other. While one can soundly criticize the 1905 paper, we should remember that it was in part a paraphrase of what was believed by the giants of the time…who got it wrong. On the other hand, in his overzealousness to insert himself into the dialog, Einstein committed grievous sins in the name of physics and his own self-promotion.

The transition from natural philosophy to physics was not complete during the fin-de-siècle period, and is still not complete, and many theoreticians have been and are philosophers and mathematical savants. While it was laudable that they used these inclinations to help them focus on the physics, they fell victim to their own abstractions and forced their ideas onto physics. In the famous case of David Hilbert, and his is not the only such example, as Einstein began to introduce tensor mathematics into physics via general relativity, Hilbert took on a tutor to help him understand enough physics for him to make some contributions in general relativity. It is this attitude that has poisoned physics in that, as Hilbert opined, physicists do not know enough mathematics to be good physicist. Thus, we have the promotion of the belief among theoreticians that the math is the physics. Such an attitude persists within theoretical physics to this day and is the source of a blight at the core of modern physics and is the issue that Frank and Gleiser address in their article "A Crisis at the Edge of Physics."

By 1906 or 1907, the 1905 paper and relativity itself were floundering and were for all practical purposes dead. Einstein could not sell it. Fortuitously, once Einstein's friend and former advisor Hermann Minkowski worked some mathematical magic on the 1905 paper, the concepts in the paper were resurrected but not the paper itself. The subsequent development of Minkowski's special relativity is generally and erroneously attributed not to Minkowski but to Einstein's 1905 paper, and Einstein was more than happy to promote that perspective. In fact, special relativity per se was resurrected only because it was bundled with the much more mathematically seductive general relativity. Minkowski died in 1909 and the direct contributions by other pioneers such as Henri Poincaré and Hendrik Lorentz were making no new progress. Lorentz ceased producing research papers in relativity at that same time and Poincaré died in 1912 after a three-year fight with cancer. Consequently, Einstein had the field essentially all to himself.

The rebirth of relativity can be attributed to Minkowski's turning the concept into a remarkably self-consistent and elegant theory that essentially relegated the mathematical style in the 1905 paper to, as modern physicists might phrase it, the junk pile of the more mundane engineering and classical mathematics. Unfortunately, when theoretical physicists made elegance its own reward, they began to ignore the underpinnings to the elegance.

Poincaré supported Minkowski's mathematics as describing a four-dimensional space with time as the fourth dimension. H. G. Wells had written his novel *The Time Machine* and other science fiction introducing novel speculations about the possibilities of new sciences and technologies. Some of these influences on the acceptance of relativity are nicely laid out in Stephen Brush's paper "Why was Relativity Accepted.?" The speculations on new physics began to catch on and when the idea of time as a fourth dimension percolated out of the scientific community, the public, and not a few scientists, became enamored to the point of obsession.

It should be noted that the four-dimensional vector space that Minkowski defined was the first baby step in moving toward the new…new to physicists…vector and tensor descriptions that Minkowski helped Einstein pioneer. Einstein recognized an innovative approach when he saw one, and tensor mechanics elevated the mathematics to a rarified level which only a few physicists of the time could navigate.

Later there will be some limited discussions relating to the final evolution of Einstein's relativity, which was the formulation of general relativity. General relativity was essentially made possible because Einstein recognized in Minkowski's four-vector a metric describing a four-dimensional space. From the first baby-step of the four-vector, Einstein molded a more general tensor description of gravity based on geometry, where the geometry was defined by the presence…or absence…of mass and a gravitational field. More importantly, Einstein had put forth what is called the equivalency principle, in which the acceleration of gravity was indistinguishable from mechanical acceleration. Though, as will be discussed, it is possible to distinguish between the two form of acceleration, but that was not something recognized by Einstein.

Something to remember is that, while there were observations driving the development of special relativity, there were no observations that motivated Einstein to produce his general theory, which incorporated his special theory. Rather, what was driving people like Poincaré and other mathematical physicists was mathematical rigor and the lack of required compliance of Newton's gravity with being Lorentz invariant.

The Lorentz transforms will be discussed in more detail in Chapter 3, but for now they are mentioned here as being a motivating factor in the development of special relativity and, subsequently, general relativity. The belief was and is that representations of physical laws, such as Newton's gravitational law, was subjected to and failed to be Lorentz transformation invariant and failed to be covariant, a tensor property thought to be required by all representations of physical laws. (This is a messy and contentious area of discussion and depends on defining what is a law and what is a consequence of the physical behavior of objects under the influence of various forces, to whit the physical manifestation of the application of the law, such as Newton's dynamic laws regarding forces and motion.) In addition, modern concepts have relegated the ideas of covariance and of contravariance obsolete. Our modern ideas of tensors and laws differ from those in Einstein's time, but many mathematical descriptions in Einstein's day were driven by the needs to meet specific tensor relationships. One consequence of relativity was to also require Lorentz invariance, which also drove the search for an alternate representation of gravity. One can see that a mathematician like Hilbert might deduce that the physicists of the time did not understand such transformation and tensor relationships well enough to formulate laws that met certain invariance or transformational requirements. Poincaré was complicit in driving the representations in physics to meet these transformation and invariance principles.

Consequently, the hunt was on for a better law of gravity, which propelled the physicists of the time to look at geometric forms for these laws. A significant characteristic of the general theory is that, in the absence of mass or in the presence of small mass, the general theory supports or reduces to the special relativity and Newtonian gravity, or so the argument goes.

Stephen Brush's paper discusses many of the reasons that relativity gained so much tractions so quickly, not only with the public but also within the scientific community. Brush was speaking to the totality of the special and general relativities. The prominent motivation was, of course, that for special relativity and as mentioned previously, experiments were being performed and results analyzed for which no known classical explanations could account for the observations. Unfortunately, Brush was a true believer, so that, even though he was a physicist, he never undertook a critical analysis of the physics that he was analyzing in his role as a history of science academic. One may again conclude that the incompleteness of the analyses of certain critical physics and our ignorance of certain classical physics abetted the misinterpretations that led to speculations about these new amazing and counter-intuitive physics that were being talked about among physicists and in the press.

Einstein opened the 1905 paper by attempting to re-derive "from first principles" what was already known and then proceeded to add what was not known. What was known were ad hoc empirical factors that explained observations, such as the Lorentz-Fitzgerald contraction, which simply worked. (The terms "contrived" or "fabricated" are synonyms for the politer term "ad hoc." So, whenever I used the term ad hoc think contrived or fabricated.) Einstein's intent was to find where these useful empirical forms came from, implying that he recognized that they were ad hoc. Both aspects of the paper relating to what was known and what was not known contain errors associated with incomplete analysis, but the newer aspects of the 1905 paper also contained unique analysis that have been essentially overlooked to this day.

We can easily see Lorentz's and Poincaré's fingerprints on the work, though Poincaré's contributions are also visible in the more mathematical descriptions of relativity associated with Minkowski's contributions. Per Marchal, those fingerprints are likely a result of Einstein allegedly having plagiarized Poincaré's early work for the 1905 paper. Lorentz's contribution was the idea of spatial contraction and of time dilation, which were consequences of the transformations that bear his name. Lorentz also developed empirical models based on the frame independence of the speed of light. And Lorentz had developed his own complete theory of relativity that was replaced by Einstein's formulation. Using all these influences, the 1905 paper was an attempt to integrate these disparate ideas and to extend the work through more first-principle detailed modeling and analysis. All in all, the 1905 paper has both very bad physics and amazing physics, but the good physics is not what is popularly promoted and taught.

And, as final note, this chapter along with the Introduction has introduced the topic of fringe science. But, along with fringe science is the notion of conspiracies, which are clearly evident in the body of modern criticism of Einstein and the so-called maneuvering of German scientists to marginalize Poincaré and his contributions to special relativity. This theme of fringe science and conspiracies will emerge in the following chapters. In a way, Einstein has provided fuel for such ideas in the way that the 1905 paper contained a propaganda style of presentation of certain topics and in Einstein's inability to clearly explain many of his ideas. This fringe characteristics will emerge further on as relativity is discussed in more detail.

Chapter 2—Background to the Original Kinematical Theory

While Einstein's 1905 paper was divided into two parts, there was really only one paper, where the first or Kinematic Part of the 1905 paper can be completely derived from the second or Electromagnetic Part. We show this in Chapter 4. One might make the case that the issue was how Einstein changed representations between the kinetic and electromagnetic aspects in the paper and proceeded to derive the same physics using a different representation. A fair point is that he did not recognize what he had done and neither has anyone else. Yet modern philosophy-of-science scholars recognize the importance of representations in how various mathematical models are developed, but these scholars rely on the work of others to show the consequences of changing representations or approaches in the middle of any given research effort.

In addition, the Electromagnetic Part of the 1905 paper contains overlooked nuggets of important physics that have no analogs within kinematic physics. These aspects of relativity are discussed in the following chapters. Ultimately, it was only Minkowski's work that salvaged relativity and it is Minkowski's approach that is taught. Only the electromagnetics and general relativity are totally unique contribution by Einstein to modern physics, but general relativity is based on a mathematical fabrication by Minkowski, and I will get to these points in Chapter 4 as the narrative in this book progresses.

During the fin-de-siècle period, many new concepts were being explored and some of the characteristics of physical kinematics were being applied to electromagnetic radiation, including to optical radiation. Radiation had both energy and momentum and the notion of radiation propagating at a fixed speed relative to any observer who may also be moving brought to the forefront the issues of reference frames and inertia. Yet, as we will show, the pioneers misinterpreted the concept of inertial reference frames, and we still do. While physicists placed certain inertial frames at infinity, they produced their physics within local frames and missed certain phenomena because they limited their perspective. And I am not going to comment on the issues with Maxwell's equations and any motivations there may have been to come up with a theory based on frames of reference. As the discussion proceeds, and for reasons that will become clear, Maxwell's equations are a red herring in the discussions, as least as far as I am concerned.

Not only were the scientists of the time motivated by analogs that existed between the kinematics of mass and the kinematics of electromagnetic radiation, there also drew the idea of waves from propagating water waves, and the idea of an aether…a mythical invisible medium…was developed to sustain the ideas that a propagating wave needed something within which to propagate. In fact, the idea of an aether is still kicked around by many physicists, both mainstream and fringe. Many experiments were performed to detect this aether, but it was never measured, and the idea was subsequently dropped though not entirely forgotten. It was the putative null results of experiments, the most famous being the Michelson-Morley (M-M) experiment first carried out in 1887, that further supplied hypotheses that motivated Lorentz, Poincaré, and Einstein in developing relativity. As with relativity itself, there are likely thousands of papers written about the M-M experiment, which is still being evaluated after 135 years.

Yet, as Cahill points out in his paper "The Michelson and Morley 1887 Experiment and the Discovery of Absolute Motion", the reported outcome of the initial experiment was apparently partially fabricated and while that has been proven, that fact is ignored and counter argued. (Cahill is one of those academic physicists whose research work has drifted into the realm of the demarcation between mainstream and fringe science, but the observational evidence driving his interpretations are legitimate observations.) Again, there was apparently no there, there in the M-M experiment, or ironically there was some there, there, which was suppressed by Michelson and Morley, or at least minimalized by them by use of the term "null-effect" (See reference Cahill 2), which was meant by M-M a measured result that was smaller than expected. Originally Cahill commented that null result was interpreted by Lorentz (and Cahill) and others as meaning nothing was observed. Though as Cahill reported initially, something was seen. If one were to believe in cause and effect, there was, therefore, one less reason to invent special relativity. This latter point is supported in Chapter 4. We will see that one observation after another that supported special relativity will be countered until there is no longer a rationale for even producing the special theory.

I used Cahill's paper as a baseline for the discussion on the M-M experiment for several reasons. One reason is that it is accessible. The second is that his papers presents a nice summary of the history of the M-M experiment. Third, Cahill is also an experimentalist, and his work is the latest and most forward thinking in that Cahill is challenging certain aspects of relativity that the community of physicists is simply ignoring. In addition, Cahill's work and references to other experiments support independent analyses in which I used a different analytical approach but arrived at similar conclusions. And, fourth, if the results of the M-M experiment are false, then the negative result no longer supports other aspects of special relativity or the concept of equivalence, which we will discuss in more detail later in this book.

Cahill is not without his critics, but many of the critical arguments fall back on nearly impenetrable arguments relating to reference frames and even use relativity to defend itself. My arguments are more first principle: I simply show that the more complex arguments are simply attempts to explain something that, as Feynman might have opined, is not known well enough to be explained. In addition, I do not agree with certain of Cahill's conclusions, which are irrelevant to his observations. The observations are what are important. Plus, while Cahill's initial observations were interpreted as undermining special relativity, in Cahill 2 he shows that he believes whole-heartedly in special relativity. Consequently, Cahill's conclusions are irrelevant to the discussion presented in this book.

Cahill concludes that the original M-M experiment was so poorly implemented and interpreted that almost nothing useful could be gleaned from it, certainly not enough to supply a foundation for relativity. In fact, Cahill remarks that the M-M data were also arbitrarily culled, something he backtracked on in Cahill 2, in order to ignore data that did show that there was some apparent motion, contrary to the published conclusions. This statement by Cahill contradicts his other statements that nothing should have been measured in the M-M experiment as it was conducted in the late 1800s. Further, Cahill claims that all subsequent efforts at using the M-M apparatus were badly conceived and executed up until he and Kitto produced a definitive model and modernization of the M-M apparatus in 2002 that, in his words, finally allowed a correct analysis of what was being measured.

It was only in the 1930s that a flaw in the original modeling of the M-M apparatus and experiment was identified, yet it remained for an additional six decades to pass until the models and experiments became reliable…per Cahill.

Some criticism of Cahill's paper relied on specific interpretations of relativity, but one cannot use a flawed theory to support another flawed theory...though that often occurs...Cahill's observations will stand on their own.

The fact is that Cahill's experimental results have been supported by other experiments and observations, though none of these efforts seem to be gaining any traction as a way casting doubt on special relativity. However, in a theme repeated over and over, those whose livelihoods depend on the truthiness of relativity ignore the critics and their criticisms. So far that approach has worked to fend off any serious attempts at an in-depth revisit into the underpinnings of special relativity.

The point is that the M-M experiment was flawed, and some conclusions were apparently fabricated or mischaracterized, and yet its false conclusions and misinterpretations were used to support the development and interpretations and continued defense of special relativity and the equivalence principle. More egregious is the fact that nothing definitive could apparently be gleaned from data collected in the original M-M experiments and that a primary stimulus for producing the 1905 paper was based on a fabrication or misinterpretation or misunderstanding, with Lorentz's complicity mostly accidental as he plied his trade as a theoretical physicist in the tried and true and accepted methods of the time. Lorentz's ideas of contraction were based on the null M-M results and Einstein's kinematics supported Lorentz's flawed interpretation, which is the underpinning to this chapter.

We can start our critique of the 1905 paper with a general error that flows through many efforts at "explaining" special relativity. Many descriptions even today of light reflecting off moving mirrors shows a component of the light's momentum in the direction of motion of the reflecting surface, what might be termed an inertial effect. Consider an atom in a vacuum emitting a photon at right angles to the atom's direction of motion. We see the photon propagating in a straight line, with the beginning of the photon's path the point at which the photon was emitted. However, if there were some component of photon momentum in the direction of motion of the emitting atom, then the photon would not be propagating at right angles to the direction of motion except as seen in the moving frame of the emitting atom. In effect, the photon's point of origin would be moving along in the same direction as the source atom. Without this momentum, anyone traveling with the atom would see the photon appear to move with a component of velocity opposite the direction of motion of the atom.

An analogy to the above inertial behavior, or the lack of it, can be given a visual metaphor. Consider a motorboat crossing a flowing river. If the boat is always pointing at 90 deg relative to the shore, the water current will push the boat sideways, so that the path of the boat is not at 90 deg but is skewed in the direction of the current, as someone on the bank of the river would observe the path. If we further think of the boat as leaving a wake behind, this wake will drift with the current. In this case, the wake is not anchored to the shore and drifts with the current. Photon trajectories and paths can be described as the wake **not** drifting with the current and, in fact, the motion of the photon is independent of the current. This metaphor cannot be pushed too far, but it highlights the fact that photon propagation is not inertial and is not like kinematic motion. Since we do not see photon propagation as having a lateral momentum component, we believe that the photon is not behaving as an inertial mass but with its own unique physics. Yet certain test-book descriptions of photon propagation insert a type of inertial behavior into its propagation path.

In the M-M experiment, a novel interferometer was built in which the incident light is reflected off a 45 deg beam-splitting mirror, with the light beam split between the transverse and longitudinal directions. There are several Wikipedia articles with good pictorial information that supports the classical description and analysis of the M-M apparatus and experiment. Descriptions of the path for the transverse beam often show that it is deflected in the direction of motion, which would only occur if the beam splitter were not at 45 deg or if the photon had a lateral (forward) momentum component. The issue is that such a deflection in the direction of motion is an inertial mass description, and photons and light do not exhibit the same inertial properties as mass.

When the photon beam is split at the 45 deg angle mirror, part of the beam has a new trajectory at right angle to the initial beam direction. The actual path of a photon is fixed in empty space as the photon defines a type of inertial path of its own in the direction of propagation that stays fixed in direction once the photon is launched and propagating in any given direction. These pathways do not "drift" in some direction related to the direction of movement of the emitting atoms or the direction of motion of the reflecting mirror. As for the apparent deflection in the moving direction, some arguments are that this is a result of Lorentz contraction on the moving mirror. Always a hypothesis that can salvage the theoretical models and descriptions.

Within an apparatus, such as the M-M interferometer, the light deflected at 90 deg out of the initial beam follows a straight path and the apparatus moves off laterally from the direction of the light path. If we have an atom emitting a photon at right angles to its direction of motion, the photon propagates off at that right angle and does not get deflected in the direction of motion simply because the atom was moving. Only the hypothetical Lense-Thirring effect in general relativity postulates such a dragging effect in a vacuum in the presence of significant mass such as a black hole. While we don't know what may be influencing the direction of emission of a photon, if a beam of photons is deflected by 90°, there is no inertial drag in the photon beam trajectory in the direction of the mirror's motion. One could say, and I do below, that this particular characteristic of photon propagation represents an inertial frame or path to which any interactions of the photon with matter can be referenced.

An experiment to measure a hypothetical inertial dragging might proceed as follows. Consider the scenario of an aircraft or satellite moving at a speed v and transmitting a radar pulse at right angles to the flight path, where the radiation is traveling at the speed c. This model uses the scenarios that the radiation or photons do not have any momentum in the direction of motion of the emitting platform. After some time t_{tar}, a return pulse is detected at the aircraft. Using straightforward geometry, we can find the time the pulse reached some target (t_{tar}) based on the total elapsed time (t_{tot}) for the return pulse to reach the aircraft: $t_{tar} = t_{tot}\sqrt{1-\beta^2}$, where $\beta = v/c$. Therefore, we find an expression that appears to contain time dilation but is only a geometrical description for the scenario.

Now, if we redo this scenario but assume that there is some photon momentum in the direction of the aircraft or satellite motion, we now find that $t_{tar} = (t_{tot}\sqrt{1-\beta^2})/2$, which is one half the value of the scenario in which there is no transverse momentum of the photons. In a terrestrial scenario, v ~ 1 km/sec and β ~ 3×10^{-6} whereas for a satellite, v ~ 10 km/sec and β is ten times larger than for the terrestrial case. We can see that while there may in principle be some observational way to determine if photons have a lateral momentum, we cannot likely do it with what amounts to radar types of scenarios. Better to perform the experiment with relativistic atoms that emit photons at right angles to the direction of motion. There is nothing in the literature that appears to address the results of such an experiment. Consequently, we take as a matter of faith that photons are not dragged along with a transverse momentum vector of the emitting atoms.

The premise behind the M-M experiment was that if the Earth was moving through a stationary aether that permeated the universe and supported the propagation of electromagnetic radiation, the motion of the Earth could be detected using the M-M apparatus. The thesis was that some subtle shift in the frequency of the radiation that is moving in the longitudinal direction should be seen as what is called a fringe shift when the reflected transverse and longitudinal radiation is recombined back at the original splitting mirror. Michelson and Morley claimed that no such fringe shift was observed or, per Cahill, it was so small that M-M referred to it as a null result. Hence, the conclusion was that a moving observe could not know they were moving. Lorentz therefore concluded that he path length in the direction of motion has shrunk, eliminating the expected fringe shift, which Cahill originally concluded that it showed that the original M-M experiment and apparatus were flawed.

The above discussion also requires some discussion over the idea of what do we mean by reference frames. Since the propagation of a beam of radiation in empty space and far from any mass is a straight line, I previously used the term inertial to describe the fact that the radiation travels in a straight path without experiencing any acceleration or deviations. Yet there is no lateral component of motion associated with the motion of an emitting atom. What happens with radiation is that the energy…and therefore the momentum…of a photon is changed by the component of motion of the emitting atom relative to the direction in which the photon is emitted while the speed of the photon is unchanged, which is clearly unique behavior unlike kinematic motion.

However, other experiments purport to show that the speed of light may not be constant. One set of observations and analyses indicate that the presence of mass can cause the speed of a photon to change relative to the radial motion of a photon in a gravitational field. Something called the Shapiro time delay, described in a Wikipedia article, purports to demonstrate such slowing. One issue with certain of these interpretations of measurements is that the models used are for the redshift of radiation. General relativity predicts that the frequency of a photon changes as a result of the radial component of motion in a gravity field, something that has been shown experimentally. But a change in frequency does not mean there is a change in velocity, yet that is how the model is interpreted. Cahill also claims that the speed of a photon is not necessarily constant. However, I have seen no models or data confirming this particular belief.

Now, how do we actually measure the existence of the radiation? The radiation must interact with matter in some way for us to even know it exists. There are only two ways matter interacts with radiation: the radiation is scattered or absorbed, but even if it is scattered, we need a subsequent observation to know that there was even any scattering. When we see light with the eye, a photon has been absorbed. When light is scattered, we can either detect...observe...the deviation of the photon by absorbing the photon with some apparatus, or we can detect a recoil of the scattering object, which is usually too small to observe other than for scattering from atomic-scale mass.

The descriptions of radiation scattering from mass requires that we know something of the speed and location of the mass relative to the trajectory of the photons or radiation flux. An experiment by Compton showed that photons act like small particles when they scatter from the particle components of matter, which are electrons, neutrons, or protons. The scattering is described using kinematic models in which the velocity (speed and direction) of the scattering particle is referenced to the trajectory of the photons. In other words, the photons are the reference frame for the motion of the particles. Therefore, a beam of radiation becomes a local reference for all interactions of that radiation with matter. If some other reference frame were defined, then the motion of the radiation and particles would have a more universal motion, yet the scattering kinematics would be the same as if only the photon trajectory was used as the reference.

The above may seem like irrelevant nitpicking, but what it says is that there is no universal reference frame when we mathematically model motion. The only reference is the relative motion that we use for describing and making a measurement, which is the local and relative inertial frame of the photon or radiation trajectory. All measurements are relative. Even when we model scattering or absorption holding the particles stationary, we use the same relative measure of the trajectory of the radiation relative to the location of the particle, which is a reciprocal representation of the geometry. In fact, while we do not go into the details, we model the scattering from moving particles by looking at the trajectory of the radiation from the perspective of a stationary particle. It is still all relative, but we can switch from one perspective to another, using the perspective that makes the mathematical modeling possible. This works because, if we are on a moving particle, we can consider the particle to be stationary and the incident radiation is transformed in a way consistent with the actual motion. This is called frame of reference switching from the laboratory into the particles frame of reference. It is all relative and reciprocal. We do not need any other reference frames. Keep this in mind as we discuss Cahill's measurement and the discussion on simultaneity and other of Einstein's thought experiments.

However, to get rid of any forward deflection in the M-M experiment and to justify a null result, the idea of contraction was introduced. Ironically, Einstein's theory of aberration supplies such a deflection without needing to consider the Lorentz contraction effect, which is discussed later. None the less, the M-M experiment reportedly did not detect an aether and Lorentz contraction was hypothesized to be one of the reasons…even though the data published with the description of the experiment did, per Cahill, show that the results were not null.

While the aether as such has never been detected, Cahill reports that motion has been measured using a more modern version of the M-M interferometer, though these finding have not apparently been sufficiently persuasive to invalidate the hypothesis of Lorentz contraction. Several prominent history of science researchers, namely Popper and Gordin, have noted that even in the absence of supporting evidence, some theories or descriptions are simply not abandoned even though they should be.

The foundations to special relativity fall into the category of false experimental and analytical justification being sustained through the century because the underlying theory of relativity has proven so useful and esthetically pleasing to physicists. Part of the reason for this persistence is that there is no viable physics to replace it in supporting the interpretations of a century's worth of observations. However, in Chapter 4, we do in fact produce a replacement which, ironically, comes from the ignored Electrodynamic Part of the 1905 paper. With a little perseverance, we can abandon the nonsense that is the Kinematical Part of relativity, which is ultimately what Chapter 3 is all about.

Continuing, the same general geometry as described for the M-M experiment is employed in certain "explanations" of special relativity in which a "photon clock" is used to show how the path of a light beam appears to a moving observer as the photons bounce back and forth between two parallel co-moving mirrors. These are called clocks because they are used to find the putative time dilation effect as an analog to the previously discussed side-looking radar scenario. In the radar model, we develop what appears to be a time dilation effect, which is simply an element of the propagation geometry.

While we proceed here assuming there is no lateral photon momentum, there are, on the other hand, ways in which a beam of light can be "dragged" in the direction of motion, but these all require the interaction of the beam with matter that is moving. The Fizeau experiment showed this to be true, but in a vacuum, there should be no such interaction. Also, the Fizeau experiment was always a longitudinal experiment in which light propagated along or opposite the flow of a liquid (water). The Fizeau experiment is described in the Wikipedia article "Fizeau Interferometer."

Interestingly, Cahill states that in modern instruments, a Fizeau fringe shift occurs when there is a stationary gas in the longitudinal path. The original Fizeau experiment had a beam of light passing down the length of a tube filled with flowing water. The beam was then deflected outside of the tube and then back to the original source and, as in the M-M experiment, there was a fringe or phase shift observed that was proportional to the flow rate of the water. Cahill repeated this experiment with his M-M type apparatus using a stationary gas filled tube. And, as Cahill reports, this experiment and results have been reported by six other experimenters.

This arrangement duplicated the Fizeau experiment because the "stationary" gas in the path is moving with the combined vector speeds of the Earth in the cosmos, an absolute speed. The speed of the photon is independent of the frame speed, so to the photon, the stationary gas is in fact flowing as in the Fizeau experiment. Combined with better radiation sources, Cahill and others have measured a Fizeau flow effect. Such definitive observations would not have been possible until the modern era. Though, that being the case, something was apparently measured in the M-M experiment, even if barely so, which must have been related to what we have been measuring now in the modern era.

To resolve all the motion components, one needs to produce an orthogonal three-axis determination of the data by re-orienting the apparatus in three dimensions, and Cahill claims to have in fact measured these orthogonal components. Cahill identifies the detected motion as absolute motion relative to the far universe, which is consistent in magnitude with the motions measured by various space-borne apparatus as they measure the ubiquitous cosmic microwave background. These velocity components are also consistent with those described from other astronomical measurements, though the magnitudes are slightly different. The key point is that the Earth orbits the sun, the sun orbits the galaxy and the galaxy has some other motion in the cosmos, and the magnitude of all these motions can be measured and have been. Therefore, we can measure absolute motion.

Cahill states that the complex vector components of motion have all been accounted for, including orientation variations that occur on a daily and annual basis associated with the above motion vectors. This was accomplished though taking data at various times of the day or year that would put his apparatus in the correct aspect to capture these velocity components. Such measurements have also been made by others.

Cahill's observations and conclusions regarding "universal" motion and his hypothesis of why we can observe these motions echoes certain points that emerged as I evaluated Einstein's model for his famous thought experiment, which is discussed next. As in the prior discussion on reference frames, we are erroneously relying on some infinitely distant reference frame in making local measurements. Brillouin raised some similar concerns about reference frames, but since he would not have been aware of the results of Cahill's detections of motion nor any other modern observation, he offered no explanations.

None the less, in the absence of the gas in his apparatus, which is in only one path, the symmetry of Cahill's apparatus is static, and, through the calibration process, no absolute measurements should be possible, only relative measurements. But the claim is that some motion was observed in the M-M experiment. The gas, however, introduces an asymmetry that is detectable. As will be shown later, by switching from a steady source of radiation to a pulsed source, motion can or should be detected without any gas in the apparatus.

Continuing, contrary to the common wisdom concerning Einstein's 1905 paper, the 1905 paper was all about optical radiation. However, the command of non-optical electromagnetic technologies was primitive at that time and little was understood about electromagnetic propagation. Parenthetically, Einstein was per Brillouin a reluctant proponent of photon theory at that time, though he used that idea in 1905 to explain the photoelectric effect, which led to the award of the Nobel Prize in 1921. (The award seems to have been premature but was rushed to address criticisms of the Nobel Committee in making a premature award to Einstein for relativity. It wasn't until 1923 that Compton showed that photons were corpuscle-like pieces of electromagnetic radiation that had both energy and momentum and behaved like particles.) Additionally, while the 1905 paper contains discussions and models for pulses of light, there was no real capacity to work with light pulses at that time of the type necessary to assure measurement accuracies for real experiments.

Continuing analyzing the 1905 paper, while that paper summarized existing concepts, it also introduced several new concepts and interpretations, one of which was the idea that an expanding spherical wave was somehow important. Poincaré had introduced a similar concept in describing what he and Lorentz were calling simultaneity. Except for a single small and ignored…but very important…model developed in the Electromagnetic Part of the 1905 paper, propagation was all linear. Einstein also tried to re-develop a series of transformations he called the Lorentz transforms named after Hendrik Lorentz who had first introduced them as ad hoc corrections and who exploited them to support the notions of spatial contraction called the Lorentz-Fitzgerald contraction. Lorentz's transforms had been developed in 1887 by Waldemar Voigt unbeknownst to Lorentz. The history of the Lorentz transforms is nicely laid out in a Wikipedia article entitled "History of the Lorentz Transforms." What the Wiki article does not emphasize is that the Lorentz transforms are ad hoc mathematical expressions that "work" as were their progenitor transforms developed by Voigt.

Einstein was never able to fully explain his 1905 approach and later dropped it from his 1916 book and essentially simply adopted the transforms as developed by Lorentz and otherwise ignored how they were derived. Despite all the machinations used to show where the Lorentz transforms come from, there is no first principle derivation of them. Where Lorentz had also simply empirically deduced the forms for the contraction, Einstein set out to find a formal way of deriving them from first principles. The issue was that neither Lorentz nor Einstein recognized that the Lorentz transforms are strictly electromagnetic and not kinematical, which we show later when we point out where these transforms do come from.

Einstein's approach for deriving the Lorentz transforms resulted in the equation $x^2 + y^2 + z^2 = c^2 t^2$, which I call a mnemonic, which was also used by Lorentz as an ad hoc tool in deriving the transforms. Einstein's approach was to determine where this equation came from and then to simply use it to find the same transforms as Lorentz had found. Einstein called the equation a description of an expanding spherical wave, whereas Lorentz did not. It is this equation that is still used to find the Lorentz transformations, though prior to this time, starting with Voigt some fifteen years earlier and on up to Lorentz, the transforms were algebraic manipulations that resulted in an "adjustment" factor that made the overall modeling mathematics work out to match observations. No one really understood their physical significance, not even Poincaré, the preeminent theoretical and mathematical physicist of the time, who introduced the notion of the Lorentz transforms forming what is known as a group, which obeys specific mathematical rules. Later, we show what the real physical significance of the Lorentz transforms is and where the formula $x^2 + y^2 + z^2 = c^2 t^2$ really originates.

However, as useful as the form $x^2 + y^2 + z^2 = c^2 t^2$ turned out to be, it is not a spherical wave description as Einstein claimed. The claim has been made that Einstein wanted to find that form and simply manipulated his approach until this equation emerged from the analysis, an assumption abetted by his inability or unwillingness to clarify what he had done to find this equation. True believers simply ignore this sordid element of the 1905 paper. Yet, many of his peers, including Lorentz, did these same types of manipulations in finding analytical expressions that made their models come out the ways that they wanted them to. After a few years, Einstein apparently began to show discontent with some descriptions in the Kinematic Part of the 1905 paper, which may have been a way of distancing himself from something that was a bit more ad hoc than he let on. Ultimately, he was rescued by Minkowski.

The notion used by Einstein in the Kinematic Part of the paper was that if we send out a pulse of light, it travels with a speed given as c and after the passage of some time, that pulse is at a distance $r = c\,t$ away from the light source. If we call the location of the light source the origin of some coordinate system, then if we have pointed the light pulse in a certain direction relative to the coordinate system, the length of a straight line or a ray from the origin to the propagating pulse is given by $\sqrt{x^2 + y^2 + z^2} = c\,t$, where the values of x, y, z depend on the specific direction the pulse is propagating and are the components of the line in a three-dimensional $\hat{x}, \hat{y}, \hat{z}$ coordinate system represented by unit vectors pointing along each direction, respectively. Thus, $x^2 + y^2 + z^2 = c^2 t^2$ is the square of a length, though as written it is the length squared of each straight line from the origin out to a distance c t., which defines a spherical surface at radius ct. Einstein only used straight-line propagation or rays in his other derivations except for one in the electrodynamic portion of the 1905 paper.

However, Einstein's comment that the form $x^2 + y^2 + z^2 = c^2 t^2$ represented a propagating spherical pulse of light was an irrelevant and throwaway observation that physicists today will argue irrationally over and in fact use as a litmus test to determine if one is too ignorant to carry on a legitimate conversation. It is simply a case of the irrational truthiness regarding physicists' willingness to believe, protect, and defend Einstein and special relativity.

So, we have the background rationale and observations…including the interpretations to the observations…that were in place as Einstein began the research for the 1905 paper. While his comments on spherical waves was first put forth in the 1905 paper, it was a foundational concept that was irrelevant to the development of the theory. The null result of the M-M experiment was an important element in Einstein's rationalization that a moving observer could not know they were moving without some outside references. In other words, an observer in an inertial (non-accelerating) black box could not know they were moving. On the other hand, a key element of the equivalence principle was that an object in freefall in a gravity field could not know it was not in an inertial state, so even though the object is accelerating, it could not know this fact without making an external observation. These, then, plus the speed of light being frame independent and all physical laws being frame independent form the underpinnings to the theory of relativity.

Chapter 3—The Original Kinematical Theory

Near the beginning of the Kinematical Part of the 1905 paper, Einstein introduces his famous "gedanken" or thought experiment, which was not uniquely his. Poincaré had introduced a similar geometry in his explanation of simultaneity, though Einstein modified it somewhat. In fact, elements of this description were first introduced nearly twenty years earlier by Michelson and Morley to describe the physics of light propagation in the interferometer Michelson designed to detect the movement of the Earth through the aether. Cahill also derived these same equations.

One of the issues with the 1905 paper was Einstein's use of clocks. By using light pulses, he was really constructing analog clocks, even though the pulses of light might imply some digital timekeeping, which did not exist in Einstein's era, even in conceptual form. Consequently, the analysis of Einstein's clocks and timekeeping may change if we were to consider modern digital timekeeping. Brillouin noted this same issue. Regardless, since we are looking at the 1905 paper, we need to use the same technology as Einstein described, so that we can understand why he may have drawn the conclusions that he did.

In the 1905 paper, and to paraphrase, an observer is located on the back end of a moving platform with a mirror located on the front of the same platform. The moving observer moves past a stationary point at which a pulse of light is emitted in the direction of motion just as the observer passes the light source. In the 1905 paper, the light pulse was emitted from a source located on the moving platform. The light pulse races to catch up with the mirror at the leading edge of the moving platform and is then reflected back to the moving observer's location. Einstein set this up with two clocks: one at the stationary observer's location and one at the moving observer's location. The clocks record when the light pulses are received at each location and the moving observer can therefore record the round-trip time of the pulse, the time from when the pulse is emitted to when the pulse was again seen by the moving observer, who sees the pulse twice. The stationary observer's clock is used to quantify the difference in the elapsed time for the pulse to make a round trip from mirror to mirror. The two clocks were used to support the notion of simultaneity, which is essentially a fabrication that we discuss later.

The round-trip time model for the pulse of light that holds for the moving observers is $t = (2\,d/c)/(1 - \beta^2)$, where c is the speed of light which is constant for both the observer and for anyone else located in any moving reference frame, where $\beta = v/c$ and where v is the relative and constant speed of the moving platform with respect to the stationary light source, d is the distance from the observer to the mirror when the observer is stationary and when v = 0, and 2 d/c is the roundtrip time for a pulse if the platform is standing still (when v = 0). This model is called a Galilean dynamic model and represents what the moving observer measures. (And yes, I said I was not going to include any mathematics in this part of the chapter, but without these simple algebraic descriptions, none of the verbal descriptions will make sense.) The equation for the round-trip time is the same result Poincaré had introduced along with the notion of simultaneity. This expression is also the one derived for the change in the path length for the arm of the M-M interferometer pointing in the direction of motion through the aether. So far, the discussion is classical propagation, but the issue was made more complicated by the researchers of the time when they introduced the idea of simultaneity.

However, the key point is that, if a measurement of the round-trip time of light propagation is made while riding on a moving platform, the time of propagation depends on the direction of propagation relative to the direction of motion of the moving platform. If no difference is measured, we would make the measurement with the apparatus turned at right angles relative to the original orientation, and if no results are measured, point the apparatus up or down relative to the plane of the first two measurements, and if no time difference is found, the platform is not moving. If any round-trip times are measured in any of the three-dimensional orientations, we can find the three-dimensional direction of motion. This is what Cahill did with his apparatus and this is what we can do with a simpler apparatus using light pulses.

In the above, we are modeling the round-trip time for the pulse from the perspective of the moving frame in which the pulse is measured twice: once to start the moving-platform clock and once to stop it. From the perspective of the stationary observer, all they can infer is the time it takes the pulse to reach the leading mirror and to be reflected. This time is $t_s = (2\,d/c)/(1 - \beta)$. Unless the moving observer triggers an additional pulse when they have received the reflected pulse, the only way the stationary observer can know the round-trip time on the moving platform is by comparing clocks after the fact.

However, the stationary observer knows that the small path-length difference exists, so they know that the time the moving observer should measure would be $t_m = (2\,d/c)/(1 - \beta^2)$. When we combine these two models, we find that the stationary observer should find that the moving clock indicates that the elapsed round-trip time as measured on the moving platform is $t_m = t_s/(1+\beta)$. This does not look like the form that we normally associate with time dilation, plus the form above depends on the sign of v, again completely unlike the accepted time dilation, which is direction independent. What has happened to arrive at this seemingly incorrect model?

One issue is that the moving observer would experience the consequences of the asymmetry in the propagation of the pulse of light. The stationary observer only sees one-half of the same propagation that the moving observer sees or measures. So, let's look a bit more closely at what the moving observer really measures, and to do that we need to look at the thought experiment in more detail.

We know that $d = c\,t$, which is that when a light pulse moving at a speed c crosses a stationary distance *d*, then the pulse has taken some time *t* to cross that distance. Consequently, *d* and *t* would be affected equally if there were some motion effects such as spatial contraction and/or time dilation. Thus, if there is some factor *f* for scaling a contraction that affects a moving *d*, then $d' = f\,d$. Consequently, $d' = c\,t'$ and $d\,f = c\,f\,t$ and, therefore, *d* always equals $c\,t$. The spatial shrinkage is called contraction whereas the shrinkage in the time is called dilation, which is the same thing as contraction. This simple logic could have been part of the origin of the idea that a moving observer cannot make a measurement that indicates that they are moving unless that measurement is made outside of their own moving frame, though there no indication that this type of logical analysis was ever carried out or reported. Combined with the null result of the M-M experiment, it could be assumed that an observer in a black box could not know they were moving by making any type of measurement.

However, the simple analysis above ignores time dilation plus spatial contraction. A moving observer experiences both dilation and contraction. Redoing the simple analysis above to include time dilation, we have that

$t'' = t'f = f^2 t$. Thus, $t f^2 = f d'/c = d f^2/c$, and $t = d/c$ as before. The point was to introduce the idea of some transformation known to or suspected by both observers which affects time and distance parameters the same way when there is motion. Inclusion of this factor shows that motion makes no difference to the measurement outcome. If a moving observer suspects that some factor f exists to affect the distance parameter, d, then they would be measuring a time, t, which is likewise being similarly affected. At this point we could argue that the above analysis does in fact prove that a moving observer cannot tell they are moving. But, there is more to it than that.

In the thought experiment, the moving observer would have a model for the round-trip time as they would measure it under the assumption that time and spatial contractions exist. Then, they would model $t' = (2 d'/c)/(1 - \beta^2)$, but, since $t' = f t$ and $d' = f d$, then $t = (2 d/c)/(1 - \beta^2)$. However, since we have time dilation effects, too, then we are measuring $t'' = f t'$ which is the same as $2 f (d'/c)/(1 - \beta^2) = 2 f^2 (d/c)/(1 - \beta^2) = t'' = t f^2$. The moving observer has recognized that there may be some effect that changes the measured values of time and distance in some way if they are moving, so the moving observer introduces the factor f just in case. The common contraction factors f cancel and the small deviation term $1/(1 - \beta^2)$ would still be present because of the inherent path asymmetry caused by motion. Consequently, the moving observer in the thought experiment can measure their motion.

Einstein also showed in his 1916 book that, when using the Lorentz transforms, which we do not describe until Chapter 4, we can show that since $d = c t$ and $d' = c t'$, defining f to be the Lorentz transforms, we find that $d' = d\sqrt{(1-\beta)/(1+\beta)}$ and that $t' = t\sqrt{(1-\beta)/(1+\beta)}$. In other words, the function f defined generically above has a specific form. The only difficulty with this is that spatial contraction and time dilation require that $f = \sqrt{1-\beta^2}$. For contraction, we have that $d' = d\sqrt{1-\beta^2}$. Therefore, we have a contradiction.

The contradiction is resolved by noting that the Lorentz transforms relate to propagation of electromagnetic radiation, so that d is a distance traveled by a pulse and not a physical object of some length d. Jumping ahead to Chapter 4, we will note that d is determined by the number of wavelengths in a propagating electromagnetic wave as the wave crosses a distance d. The factor f is the Doppler expression that relates the wavelength or frequency in a wave as measured in a moving frame. Not only did Einstein ignore the small round-trip perturbation, the ideas of physical dimension and propagation wavelengths were convolved, forcing the electromagnetic ideas into a kinematical or physical description of a length. None the less, the factor $f = \sqrt{(1-\beta)/(1+\beta)}$ will be shown to represent a Doppler in Chapter 4.

Physicist during the relativity-development era had no experience with radar or communications systems or pulsed systems and did not realize that their simple thought experiments were incomplete. What was just being introduced at the time was the idea of retarded time. Retarded time was a geometric recognition that if someone measures something in the here and now that originated elsewhere, then if they know the distance to the original event and if the speed of light is constant everywhere, then the time the event occurred where it did is given in retarded time $t' = t - R/c$, where t is the time measured by the observer and where R is the range to the original event from the location and time in which this event is measured. If an observer is also moving relative to the event location, the retarded time is modified by the motion, since the motion would affect the actual propagation distance R. As with the side-looking radar example, it is easy to convolve different events and draw false conclusions. But, as usual, it is more complicated than that.

The perspective used by Lorentz, Poincaré, and Einstein was that the moving observer cannot measure the small-time deviation based on the null M-M experimental results. Yet the small round-trip time difference that Einstein dismissed is key to the M-M experiment and others not discussed here. Therefore, Einstein's perspective was arbitrary and inconsistent from a modern perspective. True believers have argued that Einstein may not have been familiar with the M-M experiment and results, yet there is no other reason to believe that a moving observer cannot know they are moving other than by taking the results of the M-M experiment into consideration. Otherwise, where did the idea originate that a moving observe cannot know they are moving???

Also, the side-looking radar scenario modeled earlier shows how various concepts can become convolved and lead to completely fallacious conclusions. In effect, Einstein and Poincaré denied the results of the algebraic analysis on a pulse's round-trip time on philosophical grounds because the M-M experiment did not measure any motion. This observation was generalized to say that anyone moving without some reference could not know they were moving, and we maintain this viewpoint into the modern era despite all the experimental evidence to the contrary. Additionally, Lorentz contraction was introduced in an ad hoc way to account for the fact that

Michelson and Morley claimed not to have detected the aether. We can also see that while we can supposedly measure d' we must infer t' from the light propagation path. It remained until the modern era to actually measure t' in a moving clock. There is a possible exception noted later in a discussion of the rate of decay of muons produced by cosmic rays entering Earth's upper atmosphere.

Continuing, the side-looking radar scenario is just one scenario out of many in tactical environments in which a variety of signals of unknown origin are being measured by someone who may or may not be moving. We construct processes and methods of extracting information from these signals, including just detecting that a signal exists and means something. Without enough information, we have uncertainties, such as the direction and motion of the emitting source or signal. We fully recognize what we can know and when and with what certainties. We are really, in a practical sense, only dealing with propagation times and geometries and the types and qualities of information required to make some decision or prediction. We do this every day, but in the fin-de-siècle time frame, they hadn't done it and, therefore, had to invent new concepts, which they got wrong.

The idea of simultaneity was that a moving observer could not know that two events were synchronized even though the two events were synchronized relative to a stationary observer. Einstein's railroad car thought experiment described in his 1916 book was meant to show that this was true and that, therefore, there was no absolute time, only relative time. If we go back to the aircraft and two synchronized radars or communications systems, we can test Einstein's assertion. If we look at the geometry and if we are detecting two pulses separated in time, then if we have multiple aircraft moving in many directions and with various speeds, we find that the pulses from the radars are all detect with different times between the two pulses. But if we have a master observer viewing the scenario, knowing each aircraft's heading and speeds, the master observer would predict what each aircraft would see as their pulse intervals because the master observer knows where each aircraft is relative to the radars and where each aircraft is when the pulses are detected.

The issue is who has what information. In fact, by accepting he idea of retarded time, we find that if the various moving receiver platforms had sufficient information to know the range to an emitter, by using retarded time the simultaneity of the two events could be ascertained. This is what a master observer would know. The master observer has all the information they need but the aircraft do not. The scenarios from the perspective of the aircraft are starved for information.

Our modern analyses show that speed and direction are nearly irrelevant to the pulse measurements as far as propagation times and pulse arrival separations. It is easy to find that the calculated times-of-flight for the pulses in the scenarios often contain factors such as $\sqrt{1-\beta^2}$, where $\beta = v/c$. Therefore, simultaneity is an irrelevant concept and has no bearing on clock rates. It is only the starving of the scenarios of information…such as one aircraft or single pulses being detected…that causes the scenarios to become indeterminant. The concept of relative signal detection times was convolved with the idea that the clock rates are somehow being affected, yet nothing in simultaneity addresses the issue of clock rates per se, only the passage of time between pulses.

The prior algebraic analyses, however, show that the ideas of proper time and dilated time are fictitious distinctions, and the concept of simultaneity is an exercise in justifying inadequate information. There is only one time and our interpretation of our measurements can be impacted by how we interpret our measurement models. For instance, many models for time dilation are also often re-derived showing that the same time dilation can be found by simply considering spatial contraction. This is essentially what Cahill does. Or, alternatively, we can allow for the speed of light to vary with direction and not be constant, which is the same type of manipulation of the terms within an expression to force an outcome for the model. One can only know what one can know and when information is lacking, it is lacking and the conclusions as to when and where these events occurred will remain ambiguous until more information is made available.

We can go on and on modeling various scenarios adding in or deleting information. What the bottom line is that when we starve a scenario of information, we have ambiguities and complete unknowns. From a practical perspective, we either know enough or we don't to precipitate some reactions on the part of any participants in the scenarios. In other words, the clocks' relative rates are unimportant in determine what we know. What matters are changes in the relative distances and the relative speeds. Absolute times are immaterial as are the relationships between the clocks.

I am at a loss to understand how Einstein and the others mangled such a straightforward concept as simultaneity. The probable cause is their unfamiliarity with what we know as modern operations analysis, something that evolved out of the needs within WW II for both logistical and tactical actions. Simultaneity was an ignorant concept that emerged from the ignorance of the scientist at the time about how much information is needed to successfully correlate events. These issues are ongoing practical and tactical issues that we have learned to handle in a multitude of ways, including concluding that sometimes we don't have enough information available to initiate some reaction. It is all about information, and in most of Einstein's thought experiments, he set them up so that there was not enough information available to reach any actionable conclusions. But, just a likely, by using a time interval to represent time rates, the narrative was supported that time was relative. Once the idea took hold that time was relative, other interpretations were dismissed by what Pagan Kennedy refers to *déformation professionnelle* in motivating the fin-de-siècle physicists to look though what she also calls a knothole, where *déformation professionnelle* means that some particular perspective is locked in. This perspective also likely indicates the initial formation of a consensus, which, over time, becomes a paradigm which further reinforces the *déformation professionnelle* behavior pattern.

The logic of simultaneity is further contradicted by practical measurement processes. For some reason, Einstein only used single light pulses. There was no reason for this. Nothing drives the derivation this way. With a few more measurements, we get rid of certain ambiguities, just like the use of multiple digital clocks allows us to recognize and measure a persistent round-trip time perturbation as being related to the motion of the object creating asymmetrical light paths.

It is likely that intellectually, the idea of the speed of light being constant for all moving observes was confusing to the physicists of the time. It is easy to make comparisons with kinematics, which is what Einstein attempted in the first part of the 1905 paper. In a billiard game on a moving train, for instance, the billiard kinematics does not show any influence from the fact that, in the absence of friction, the billiard table is moving. That result carried over into the thought experiment might lead one to conclude that the moving observe cannot detect their motion. Such observations with billiards also lead to another false association, which is to attribute a component of photon momentum along the direction of propagation as an analog to inertial motion.

The next step in the mathematical logic is to ask, how can Minkowski's reformulation be correct if it predicts something that is wrong? The answer is that it cannot be correct. Even the effects it predicts that are not predicted within Einstein's 1905 paper and despite how useful those other models may be, e.g., the relativistic kinetic energy, they must also be false. Nothing about Minkowski's four-vector can be true. We simply cannot keep any part of it intact. This conclusion is also reiterated in Chapter 4 where we discuss how and why Minkowski introduced his reformulation.

More puzzling, however, is the unwillingness of modern physicists to re-examine many of these fundamental postulates and ideas that motivated Einstein in his development of relativity. From a practical perspective, physicists are fully capable of performing such analyses and coming to the same conclusions that I have. The real issue is that they will not do it as a matter of the principle based on the sociology of modern academic physics.

At this point, we could drop any further analysis of the 1905 paper, since the underpinnings to relativity and any early motivations to pursue a theory of relativity were completely wrong. However, the 1905 paper just keeps on giving, both bad physics and some uniquely powerful physics. We mentioned the Doppler and aberration effects in the previous paragraphs, and these are part of the unique contributions that Einstein's 1905 paper presented, and it will be shown that these effects are in fact what relativity, such as it is, is all about. Consequently, we need to forge on into the electrodynamic part of the paper.

However, and this is a big however, as Brillouin pointed out and as mentioned previously, there is a problem with Einstein's thought experiment and his clocks. They are analog. While the prior conclusions are valid, some caveats need to be inserted into the discussion. In so far as we have certain unacknowledged physics associated with light pulses bouncing back and forth, there are many technologies that have such photon behavior, such as lasers, masers, and atomic clocks. These technologies are, therefore, susceptible to the photon round-trip perturbations discussed in this chapter.

But what happens when we have digital clocks based on different physical principles? The simple thought experiment looks different when digital clocks are used. If we have a digital clock with no photon propagation, then we could encode a signal from the clocks that is continuously broadcast to a non-moving external observer. While the digital signal would be Doppler shifted because of relative speed, the content of the signal would not change and would only be up or down shifted in its spectral content, but the information would remain the same. If the thought experiment is repeated using digital clocks and digital broadcasting, we could monitor any system's time latencies while monitoring the thought experiment in real time. If there is time dilation, the digital clocks would run slower and the small roundtrip pulse perturbation would still be there but would not be consistent with our modeling of what it should be as we monitor the speed of the platform upon which the experiment is being conducted. What the above implies is that we are not conducting the proper experiments to either validate or reject time dilation of some type. If we observe time dilation directly, we would then have to understand why it occurs, since from the perspective of Einstein's 1905 paper, there is no time dilation.

We alluded to one measurement that seems to support time dilation directly, and this is what is known as muon decay. When high-energy cosmic rays hit the Earth's atmosphere, high-energy muons are produced. Muons have a decay half-life of ~2.2 microseconds. Given that the muon speeds are close to the speed of light, the muons are predicted to have a flux of undecayed muons of a certain magnitude at the surface of the Earth. However, there are more muons reaching the surface of the Earth than can be accounted for without invoking time dilation, which is supposedly slowing the internal clock of the muons and delaying their decay.

Many of the measurements on muon decay and fluxes were made more than fifty years ago and it may be that these experiments need to be revisited in terms of modern modeling improvements and better knowledge of the cosmic ray fluxes and penetration rates into the atmosphere. One concern is that the modeling is so complex that, as is common, short-cuts and heuristics were used for the models and never re-vetted for their accuracies. It may be that the time dilation is an artifact of the assumptions used to develop the models used in estimating the expected flux from the cosmic sources versus laboratory sources of muons. For instance, the impact of multiple scattering events in the forward direction was not recognized until forty years ago. In various experiments shining lasers or observing sources when aerosols such as fog is introduced between a source and receiver, the magnitude of the detected signal does not decrease according to linear scattering laws, often referred to as the Beers-Lambert extinction coefficient. If the fog particles scatter losslessly, the signal no longer decreases. Such contradictory effects were unknown at the time the muon decay experiments were being made and could be one source of falsification of the measurement of the muon decay and subsequent measured flux. Using the information in the 1905 paper, we cannot predict time dilation or length contraction, which is not to say that they do not exist in some form for certain scenarios, but not for Einstein's scenarios.

Chapter 4—The Original Electromagnetics Theory

To carry the discussion of relativity any further, we need to discuss specific models developed in the second or Electrodynamic Part of the 1905 paper. In the second part, Einstein used the Lorentz transforms he supposedly found in the first part. This chapter is important because in it we identify what relativity really is and supplies an alternative to the kinematic models that were falsely derived in the Kinematic Part of the 1905 paper and, consequently, we are able to identify where all the modeling efforts in the first and Kinematic Part of the 1905 paper really come from.

Not everything in the Electromagnetic Part of the 1905 paper will be discussed. The discussions will be restricted to showing how the Kinematic Part of the 1905 paper is wrong at every level. We can do this through algebraic discussions, though to get to the algebra, some more advanced concepts will be introduced. These concepts, however, are part of the more advanced basic physics that is taught in the first few years at universities in both calculus and physics, so the concepts will be simply stated without any further discussion or description.

One caveat for the rest of the book is that the effects being modeled are macro effects, which ignores the true physics of photon or electromagnetic interactions at the quantum level or with nanoparticles, in which polarization plays a strong role in the interaction, whatever polarization may be when discussing photons. In empty space, there are only photons, whatever those are. In matter, however, things change, and the quantum effects described by field theories begin to emerge. Without a doubt, there are properties of photons that complicate how they interact with matter, but by and large, we can focus on the macro-effects that are essentially classical and, from a practical perspective, we stay with the radiometric and optical discussions as the book proceeds.

In the 1905 paper, the Lorentz transformations were purported to have been found in the kinematic portion of the paper and were then applied to the propagating part of a plane electromagnetic wave. Propagation is analytically described using an exponential form that can also be stated as sine and cosine functions, which also describe the supposed oscillations of the transverse components of a propagating electromagnetic wave and as defined by Maxwell's equations. Stated as an exponential, the propagation in the x-direction is $\text{Exp}[i(\omega t - 2\pi x/\lambda)]$, where x is a location on the x-axis, λ is the wavelength, ω is the circular frequency and is $\omega = 2\pi\nu$, where ν is the frequency, t is time, and "i" is an imaginary number. In empty space, $c = \lambda\nu$ and c is the speed of light. If we look at a given point moving with the propagating wave, we have that $i(2\pi x/\lambda - 2\pi\nu t) = 0$. Making further simplifications, we find that the above expression, which is called the complex phase of the propagating wave, becomes $i(x - \lambda\nu t) = 0$ or $x - ct = 0$. This form looks like a kinematic motion expression and should start to form the idea that electromagnetic propagation at a certain level uses equations that look like kinematic motion.

In the 1920s, with the development of quantum mechanics, the fundamental quantum mechanical equation was Schrodinger's equation, whose solution was a wave function represented as an exponential. Hence, the term wave mechanics for the early quantum mechanics. In the exponential representation, the momentum of a particle was dependent on the velocity of the particle. In the same way as the complex phase represents a propagating photon, the complex phase of the exponential in quantum mechanics contains the velocity of the particle, and the location of the propagating particle was determined by setting the complex phase equal to zero.

When we look at the algebra for an object moving in the x-direction, we have $x = vt$, which is the typical algebraic and Newtonian expression for motion. We do not refer to it as propagation, but since $x = vt$ is recognized from the complex phase, then the object is propagating along the x-axis and is located at $x - vt = 0$. The propagation of a particle will be revisited later, but it suffices here to recognize that propagation can have several representations, but Einstein would not have been familiar with the propagation of a particle as being represented by an exponential function, but he would have recognized that a light pulse could have both representations, one being similar to a kinematical Newtonian representation.

If we look at the linearly propagating light pulses within the context of the thought experiment, we would develop a representation based on the exponential to describe the light pulse's propagation. Then if we let the pulse propagate to the mirror and then be reflected, we would have to incorporate the consequences of the pulse being reflected, which not only changes the direction, it changes the wavelength and frequency of the light pulse. If we just look at the consequences of the behavior of the complex phase throughout the thought experiment, we would clearly see that the descriptions look exactly like the kinematic propagation representations. But, in addition, we would see that the propagation amounts to counting wavelengths for the outbound and then reflected pulse. If we compare these pulse descriptions with comparable propagating pulses traveling in both directions but not experiencing a reflection, we would relate these pulses via the Lorentz transforms. While the Lorentz transforms are identified with propagating plane waves, they also hold for pulses, which are superpositions of many plane waves that form the pulses. It will become clear that Einstein mixed his representations of propagating light pulses without recognizing what he was doing.

The argument goes that when two observers are moving relative to one another along the same line, at the moment when they are coincident and passing each other, they are looking at the same point on the propagating wave. What they each sees or measures is related by the Lorentz transforms. Analytically, the primed parameters in the following expressions are in the moving frame and must be equal to the same expression in the stationary or unprimed frame. In other words, a plane electromagnetic wave looks like a plane wave no matter what the relative speed of the observers at the point where the wave is measured. Simplifying as before, $x - ct = x' - ct' = 0$. If we note that $x = ct$, then $x^2 = c^2 t^2$. This would be recognized as the origin of the manipulations Einstein made in the first part of the 1905 paper used to find $x^2 + y^2 + z^2 = c^2 t^2$, which for motion along the x-axis becomes $x^2 = c^2 t^2$. The reader should be starting to get the idea that Einstein's "kinematics" was nothing of the sort.

Using the notion of the complex phase, we might form the above model in a different way. If we have a pulse emanating from a stationary source that is directed toward the moving platform with a mirror attached to the front of the platform, we would transmit a pulse from the rear of the moving platform just as the stationary pulse arrives. We then have the two pulses propagating parallel to one another. If some analog to Maxwell's demon is riding on each pulse, each demon would look at the other's pulse and compare what they are seeing. Since the two pulses are both propagating at c, the demons stay side by side. Each would look at the physical characteristics of the other's pulse, and what they measure would be related by the Lorentz transforms. The pulse from the moving platform would have its wavelength, frequency, and time parameters labeled with a prime. This sets the stage for how the Lorentz transforms are employed.

Taking the discussion further, the Lorentz transforms are found by noting that these transforms are orthogonal, where orthogonal means the coordinates are independent, and that $x' = v\,t'$, where v is the speed of the moving observer or frame along the x-axis relative to any speed the unprimed observer may have, which is taken to be zero. The transforms are functions of v, and the transforms that solve $x - c\,t = x' - c\,t'$ is the same as the transforms that solve $x^2 - c^2 t^2 = x'^2 - c^2 t'^2$. Consequently, the form of the transforms that Lorentz found empirically and that Einstein attempted to find analytically has a physical significance that has never been noted. As a result, the entire process of finding the Lorentz transforms by Lorentz and Einstein were mathematical manipulations to find appropriate algebraic expressions that the physicists wanted to find to achieve their goals and fulfil their intellectual agendas.

Had physicists at any time recognized the origin of the expression, they would have recognized the association between the complex phase of a propagating electromagnetic wave and the quasi-kinematic manipulations from the Kinematic Part of the 1905 paper. However, Einstein could never explain how his mathematical manipulations in the Kinematical Part of the 1905 paper actually arrived at the mnemonic that allowed the Lorentz transforms to be found. We can conclude that no part of the analyses in the Kinematic Part of the paper is correct or has any significance. The thought experiment was simply a kinematical-like description of the complex phase of a propagating pulse described in a moving frame. In short, there was nothing kinematical about the kinematical part of the 1905 paper except that an observer moves at a constant speed in a straight line. The real meat of the 1905 paper is in the Electrodynamic Part of that paper.

Einstein used the complex propagation phases to find the Doppler shift formula and another effect known as aberration by using the Lorentz transforms that he supposedly found from his putative kinematics developed in the first part of the 1905 paper. Neither the Doppler effect nor the aberration has been fully described or exploited, but the Doppler model is used in all electromagnetic systems for which Doppler information is gathered, measured, and used to ascertain the speed of an object. When we use the Lorentz transforms to correlate the primed or moving parameters with the unprimed or stationary parameters, we find that we have multiple simple equations in the x, p, t, and $E = h\nu$ parameters. We are simultaneously solving these equations, and the Doppler is the expression that relates ν with ν'.

Still, we also have another variable missing from the above equation, since in the above complex phase, we have limited propagation to being along the x-axis. If we allow the path to be more general, we have another parameter φ which is the angle between the direction of propagation of the radiation and the direction of motion of the moving observer. This completes the set of simultaneous equations that allows each side of the phase to be zero at the same point in space and time for any relative motions.

The models arose from a brilliant observation and subsequent mathematical representation and manipulation made by Einstein that resulted in both a Doppler expression that is unique and another expression that was even more unique, which is the model for the resulting φ', which is called an aberration angle. Lorentz almost reached this same point, but Einstein's approach was the one that worked in identifying the Doppler and aberration expressions.

The Doppler expression was immediately used, but not so the aberration, whose true role is still unrecognized, which is discussed in subsequent chapters. Part of the lack of recognition is that the physical origin of the aberration is not understood. Many physicists and astronomers simply view it as occurring when a moving measurement system is performing a measurement without there needing to be a physical reason for it other than motion. The aberration occurs because there is a real interaction of the radiation within a moving surface at the atomic level, which causes an energy and momentum change in radiation, which leads to an angle change in the propagation direction, called the aberration. Many texts claim that the aberration is simply a result of relative motion, but it is more than that.

Einstein's Doppler and aberration models are well known. We analyze the aberration in the following chapters, but we will use the Doppler model in this chapter. The Doppler model is given as $v' = v(1 - \beta\cos\phi)/\sqrt{(1-\beta^2)}$, where the frequency incident on a moving object is v and the object is moving at some angle φ relative to the direction of propagation of the incident radiation flux. The speed of the object is $v = c\beta$ and the Doppler shifted frequency is given as v'. If an object is moving into the radiation flux, $\varphi = 180$ deg and the Doppler model becomes $v' = v\sqrt{(1+\beta)/(1-\beta)}$, which we use in this chapter. This model has been validated, at least for low and non-relativistic speeds, which are typically speeds below ~ 0.1 c.

The factor $\sqrt{(1+\beta)/(1-\beta)}$ can also be recognized from Chapter 3 when we found that one form for the scaling parameter f was $\sqrt{(1+\beta)/(1-\beta)}$, which related both d and t in the primed and unprimed systems: $d' = f\,d$ and $t' = f\,t$. Since the geometry was linear in the thought experiment in which we were finding d and t, we found that f was the linear Doppler model shown above. How is this possible? How do we Doppler shift time and a length? Is this telling us that time dilation and spatial contraction are real but are related by a different multiplier than the accepted value?

The above issue is resolved by going back to the beginning of this chapter when we defined the complex phase as $\mathrm{Exp}[i(\omega t - 2\pi x/\lambda)]$, which can be re-written as $\mathrm{Exp}[\psi]$. The symbol ψ is the complex phase, which we expanded to look like $x - c\,t = 0$. The same quantity in the moving frame is ψ', and we find the Lorentz transforms by solving the equation $\psi = \psi'$. For the linearized model, we have $\psi' = f\,\psi$, where $f = \sqrt{(1+\beta)/(1-\beta)}$.

Another way of looking at this is to think of a wave as a real thing with some oscillating features. A wave has a wavelength and the size of the Doppler shifted wavelength is different from the non-Doppler shifted wavelength. The time must be shifted the same way, so that the speed of the propagating wave is c. Therefore, this again reinforces the false notion that the idea of propagation over a distance was equivalent to a physical length. In reality, all we are doing is counting wave lengths or oscillations in a propagating electromagnetic wave, and when a wave is Doppler shifted, the counting changes. However, the above is a result of identifying radiation as a wave and not as a flux of photons, and we address this issue later.

Whether these Doppler expressions prove to be heuristics or first-order approximations is yet to be determined, but all indications are that they are valid models for all speeds, which is discussed elsewhere in this book. However, as has been pointed out previously, Einstein's models are not developed with measurements in mind and Einstein applied them without trying to fit them within a measurement scenario. In addition, the caveat about these models being first-order is echoed by Brillouin, though he had other reasons than I did for making the same observation. None the less, we will argue later in the book that the way we use the Doppler likely means that it is correct for all relative speeds of objects.

What the Doppler model indicates is that when a measurement is made from a moving platform, the measured frequency is shifted to ν'. The reason for this shift is the Doppler effect. The actual Doppler shift is $\nu' - \nu = \Delta\nu$. In everything that follows, the intention is to describe the shift and not simply the new frequency, and the term shift is often left off and we simply refer to the Doppler.

Brillouin points out that Einstein's Doppler model implies that the measurement apparatus has a large momentum and mass compared to the energy and momentum of the photon's being measured. He further points out that recoil during the interaction is omitted from the discussions. We speak to this omission in the next few chapters and show that it is really an inconsequential issue, since the Doppler is used to find the proper frequency to use in modeling a collision, and it is during a collision that the recoil is important, as it is during the emission of a photon. Einstein, on the other hand, simply assumed an incident monochromatic flux without regard to how it originated or how it is measured.

As an example of the types of assumptions that Einstein made about some of the models he developed, we can investigate something Einstein claimed he developed in the 1905 paper, which is the addition of velocities. Einstein's approach was to simply apply the Lorentz transforms directly onto a moving object and show that if we directly add two velocities, the resulting sum is not a simple Galilean addition. The resulting velocity was a complex function that seemed to indicate that no matter what the speeds of the two objects are, the sum of their speeds will always be less than the direct sum. In other words, the relative velocities of two object with respect to one another can never exceed the speed of light. Furthermore, if the two velocities approach the speed of light, the sum also approaches the speed of light! We find such a sum from a scenario such as an aircraft firing a missile at some target. The standard or practical approach would be to simply add the velocity of the missile to that of the aircraft. However, the addition of velocities gives a modified result, which is insignificant under practical scenarios such as the missile-launch scenario.

None the less, since the Lorentz transforms were derived using electromagnetic propagation, it stands to reason that some comparable transforms should exist for a propagating mass, which are fundamental to quantum mechanics. The quantum descriptions also use an exponential form in describing the propagation of a physical mass. The first notion, then, is that we can try to find the equivalent transforms for a mass propagating at some speed less than c. Remember that the Lorentz transforms were with respect to one observer moving relative to another but both simultaneously measuring an electromagnetic wave traveling at the speed of light.

When we carry out the same type of derivation using propagation speeds less than the speed of light, we have a problem. Such sub-light propagation is a common model used in quantum mechanics. One might assume that propagation is propagation, whether it be a mass or a photon. However, it turns out that the approach that supplies the Lorentz transforms for electromagnetic radiation cannot be used for any mass propagation speeds other than c, c/2, and zero. The comparable transforms for mass motion are otherwise mathematically undefined. Hence, the application of the Lorentz transforms directly on propagating mass would seem to be invalid. Yet, this is essentially how Einstein arrived at his addition of velocities model…it is invalid.

In quantum mechanics, we have a physical mass moving at a velocity in some direction. While the linear Galilean transforms work for kinematics, they did not work for electromagnetic propagation. Consequently, the Lorentz transforms are unique to photon propagation. The Schrödinger wave equation was made invariant…the same in any propagating frame…by de Broglie's hypothesis that the wavelength of a particle could be defined as $\lambda_{particle} = h/p$, where h is Planck's constant and p is the momentum of some mass. This eliminated the need for the Lorentz transforms in the Schrödinger equation. Still, there is no rationale for using the Lorentz transforms on propagating mass. One could say that the success of quantum mechanics was de facto proof that Einstein's addition of velocities was wrong, but if so, what is the addition of velocities?

The scenario used to find the addition of velocities is that one moving observer is measuring the speed of a second moving observer. The question is, how is that measurement occurring? The addition of velocities as defined by Einstein was simply the application of the Lorentz transforms on a moving mass. Einstein looked at a mass in a moving frame and then looked at the same mass in a frame moving relative to the original moving frame, thereby setting up the scenarios of seeing what the combined or added velocities would look like.

The real issue is, how do we know how fast something is moving? To answer this question, we can look at the addition of velocities using different scenarios, some based on recognizable radar scenarios and others based on thought experiments like that used by Einstein. The key is that we need to have some way of measuring speeds and not to just assume what the speeds are without measuring them. However, to construct these scenarios, we will need to enter the realm of how we actually make practical measurements.

Let's set up a scenario in which we are modeling a moving radar and a stationary retroreflector. The radar simply emits pulses at time increments Δt, and the arrival time of each pulse at the retroreflector is then given by the decrease (or increase) in the path length and pulse travel time, so that the arrival of the pulses at the stationary retroreflector would have a time of arrival per pulse that is different compared to the pulses emitted from a stationary radar. For the radar approaching a stationary receiver or target or retroreflector, each pulse's arrival time is shorter than for a stationary emitter. Using a Fourier analysis of the arriving pulse stream between the two scenarios, which is comparing the frequency from a stationary and from a moving radar, we would calculate that change in arrival times is equivalent to a frequency upshift. Simplistically, the change in each pulse's arrival interval is given by $\beta_{radar} \Delta t$, which can be positive or negative, depending on whether the emitting radar is moving toward or away from the retroreflector, or better, if the transmitter is moving in the direction of the transmitted radiation. We determine the sign for β depending on whether the emitting or reflecting object is moving with or against the direction of radiation propagation. If we have a change in arrival time Δt, the equivalent change in frequency is $\Delta v \sim 1/\Delta t$.

Now, we give the retroreflector a velocity of its own. The arrival time interval at a moving retroreflector would then be subjected to a further modification associated with any motion of the moving retroreflector, and we would have $\beta_{retro} \beta_{radar} \Delta t$ as the modified pulse arrival times when both the radar and the retroreflector are moving. We can preserve the relative directions of motion by the sign on the respective values of β.

The first measurement occurs with the emitting system moving at some velocity v_1 toward and relative to the retroreflector and emitting a signal of frequency v_1. The measured Doppler frequency at the stationary retroreflector is given as $v_1' = v\sqrt{(1-\beta_1)/(1+\beta_1)}$, where $\beta_1 = v_1/c$. We then consider the speed the retroreflector is moving relative to the flux propagation direction. The first Doppler frequency incident on the retroreflector is now itself subjected to the Doppler associated with the independent movement of the retroreflector, which we call v_2. We now have a new frequency $v_2' = v_1\sqrt{(1-\beta_2)/(1+\beta_2)} = v\sqrt{(1-\beta_1)/(1+\beta_1)}\sqrt{(1-\beta_2)/(1+\beta_2)}$. The new Doppler frequency would also be equal to $v_3' = v\sqrt{(1-\beta_3)/(1+\beta_3)}$, where β_3 defines a new equivalent velocity that gives the same Doppler as the two sequential Dopplers associated with the velocity increments v_1 and v_2, where the sign of v is determine just as we determine the sign of β. Solving for $β_3$ we find that $v_3 = (v_1+v_2)/(1+v_1 v_2/c^2)$, or, in terms of $β_3$ we have $β_3 = (β_1+β_2)/(1+β_1 β_2)$. These are the results Einstein derived for the addition of velocities. It is necessary to remember that in the case of v_1, the radiation is emitted in the same direction as v_1, whereas, in the case of v_2, if the motion of the second object is in the direction of the first object, the velocity v_2 is toward the source of radiation. Both motions preserve the positive sign relative to the direction of motion with regard to how each sees the radiation pulse.

The scenario is often set up as a stationary observer measures the speed of an aircraft that is receding from a radar. The receding aircraft then fires a missile or gun in the same direction. The question is posed: if the stationary radar measures the two objects receding at relative speeds v_1 and v_2, what speed would the first moving object measure as the speed of the second object that is receding from the first object? The only way to do this is with a radar. This scenario gives the same results as derived previously.

But what is this form for $β_3$ telling us? In Einstein's approach to the addition of velocities, Einstein did not actually find a Doppler shift directly though he was inferring it via his use of the distinct Lorentz transform in his scenario. He states that in a moving frame he has a particle moving with respect to his moving frame at some velocity. He then wants to know how fast the particle is moving relative to the frame from which his initial moving frame was referenced. He simply tells us what the speeds are and proceeds to perform a velocity calculation using the Lorentz-transformed distance and time parameter associated with each frame. From this he finds a relative velocity and calls it the composition of velocities. In the radar and retroreflector scenario, we measure the speeds using the Doppler shifts directly, just as we would in our customary practices. We picked the speeds v_1 and v_2 as arbitrary speeds, but we confirmed these velocities via direct measurements. The addition of velocities is not a limitation in kinematics and it is not really an addition, but it is a result of the characteristics of the Doppler shifts for EM radiation.

We can develop other scenarios, but the consequence is that it is the measurement process using EM radiation that creates the perceived limitations associated with the addition of velocities. It is not the velocities that are being added, it is the Doppler shifts associated with the relative motion of one or more objects for which we are finding the relative motions via Doppler measurements. We must have a reason to simply apply the Lorentz transforms, and the only valid reason is to find the Doppler shifts so that we can find the relative velocities. As developed by Einstein, the composition of velocities is another mathematical manipulation.

Under normal practices, we do not really find an addition of velocities, though we can. One could create various aircraft tactical scenarios in which a ground-based controller would want to keep aircraft one appraised of the state of aircraft two in the case that the information that aircraft one has about aircraft two is incomplete or perhaps wrong. This would be common in certain tactical aircraft traffic control scenarios. Such scenarios also could occur in modeling the active guidance of some moving object, such as a missile relative to the launching aircraft. In general, in physics, the addition of velocities is simply an algebraic manipulation. And so it goes.

We will need some information supplied in the next chapter to continue these scenarios. Experimental evidence shows that moving emitting atoms do impart a Doppler on the emitted photons. Therefore, since the initial frequency from a moving radar is reflected back to the moving radar…a typical scenario, we need, therefore, to understand how the moving radar affects the Doppler of the reflected signal. It may cancel the initial Doppler shift or it may not. The Doppler function is non-linear, so we need to understand how the actual models respond to a three-Doppler-sequence process.

We can continue the process begun when we found v'_2, which we redefined as v'_3 when we identified the equivalent factor β_3. From this observation alone, we know that the final measured v'_4 back at the emitting radar will define a new equivalent value of the combined velocities called β_4, which is a result of the non-linearity of the Doppler function. For typical closing velocities ~ 2.0km/sec, which is each aircraft moving at 1.0km/sec toward the other, the correction function is $1/(1+\beta_1 \beta_2)$. If the radar is emitting in the forward direction, the Doppler is upshifted. If the retro-reflector or target is approaching the radar, the Doppler is again upshifted. Consequently, the received retro-reflected radiation is again upshifted.

In the normal non-relativistic cases, $v<<c$ and the addition of velocities is linear. Two objects approaching one another have a closing speed that is the sum of the two speeds. The correction factor is very small and is typically ignored. How small is it? For the case of the two speeds being 1.0km/sec toward one another, we have a correction factor ~ -10^{-12}, which is unmeasurable. For a science-fiction scenario in which the closing speeds are 0.2c, we have a measured closing speed of 0.198c rather than 0.2c. From a

practical perspective, we can ignore the relativistic addition of velocities, at least for tactical scenarios for the next century...maybe. And again, remember that the correction factor is a result of Doppler measurements, and the individual speeds are still v_1 and v_2.

However, it remains to be seen if we ever truly need the addition of velocities. In a particle accelerator, we would have to contrive a scenario in which truly relativistic objects are closing at near light speeds and emitting radiation for which some Doppler shift may need to be calculated. In the above radar scenario, we had a return signal to consider. For particles, only if a particle decays with a gamma ray which strikes an oncoming particle which scatters that gamma ray back toward the original emitting particle do we emulate the radar scenario.

In a linear radar scenario, the Doppler upshift of the initial emitted photons occurs twice, once during the initial transmission and again upon receipt of the retro-reflected photons. However, this second upshift is the only one that is measured. A linear block diagram of a radar transmitter-receiver shows that the photons out of the microwave source is already Doppler upshifted before the output beam is split between the antenna and the receiver mixer. Thus, the total measured Doppler is defined just by the incoming retro-reflector speed and the relative transmitter speed. While we have three Doppler shifts to account for, only two are measured by the radar receiver. We can see that the correction factor might now be given by $1/(1+\beta_1 \beta_{2(} \beta_3)$, though the way the initial Doppler shift is used in detection means that we do not actually need the factor with three values of β. If we did need to include an additional factor of β, the discussion shows that it makes the apparent deviation from simple addition even smaller than for just two values of β.

If we use another science fiction scenario of two craft traveling toward one-another at 0.99c, then we find that the final measured Doppler computes a closing speed of ~0.99995c rather than 1.98c, which is a substantial error. And if the closing speeds were each 1.0c, we compute a Doppler closing speed of the two toward one another of 1.0c and the sum of their closing velocities would appear to be the speed of light. Consequently, Einstein's addition of velocities was inadvertently finding the impact of a practical Doppler measurement in finding the closing speeds while not calculating the actual closing speeds.

The larger issue is, therefore, how we interpret what we measure or can measure. The above scenarios do show that what we measure depends on what our models tell us is happening. In a measurement, what we interpret the measurements to mean depends on how well we understand the model we use to describe what we think is occurring. We can now look more closely at what the Doppler expression is telling us.

Chapter 5—The Doppler and Radiation Pressure

The derivation of the Doppler and aberration is found in many texts, though without detailed explanation and only about the mathematical manipulations. The physical meaning is never included nor is the observation that the Lorentz transforms can be found from the same descriptions that yield the Doppler and aberration models. And, no one appears to have looked at the derivation as a description or model of a physical process. I have an objection to the "wave equation" approach to the derivations in lieu of the photon flux approach, and the photon flux approach is discussed in Appendix 2, but to be clear, the two approaches are equivalent.

In addition, there appears to be an egregious mathematical error that has persisted for a hundred years in using the Doppler and aberration models. While the error is mathematically egregious, it is, from a practical perspective, irrelevant and, therefore, is the reason why it was never noticed. This will be discussed later in the chapter after we investigate how the Doppler has been under-utilized, which, ironically, would not have happened if people had read and understood the Electromagnetic Part of the 1905 paper.

To recognize how the Doppler has been under-utilized, we need to again describe how we make a measurement. The descriptions of the complex phase as representing a flux of photons is a starting point for understanding that the flux of photons that are being measured supply more than a "signal" out of the measurement apparatus. Photons have momentum and whenever a photon strikes an object there is momentum exchanged with the object. Therefore, the photons also supply a radiation pressure on the objects they strike, which leads us to Einstein's radiation pressure model from the 1905 paper. Ever heard of that? It is there and totally overlooked and forgotten.

As a side and cautionary note, I asked this question on the Quora website, which is one of the "recommended" sites to ask technical questions that are answered by "experts". The response I received was, to paraphrase, "Why should anybody care about Einstein's radiation pressure model? We already have perfectly good radiation pressure models, many developed before

Einstein?" The response stunned me as did the lack of any pushback from other professionals who would be familiar with the concept of radiation pressure. I have concluded that the answers posted on Quora are, on the one hand, either directly out of books or Wikipedia or, on the other hand, are anecdotal and first-hand experiences, which are often worthwhile and sometimes very interesting and enlightening, but anything scholarly is highly suspect and certainly not innovative.

Einstein's radiation pressure model has been totally ignored, and while his model is inaccurate, it contains one feature that is completely lacking in modern models of radiation pressure. The missing feature is the Doppler shift. We described the radiation pressure in much more detail in Appendix 1. However, radiation pressure on objects such as solar sails or spacecraft, in which some source of radiation, usually from the sun, applies a pressure to a broad surface and produces propulsive forces, which requires the incorporation within the radiation pressure model of dynamic Doppler factors that are completely ignored.

Radiation pressure is mostly confined to discussions of space objects, though it plays a substantial role in nuclear explosions, laboratory simulations of nuclear explosions using laser compression, plasma devices and, as we will see, a hidden role within particle accelerators. There was only one reference in the solar and radiation sail literature on the role of the Doppler, and this role was dismissed by noting that Doppler effects…or any relativistic effects…are considered inconsequential for non-relativistic speed ranges. This viewpoint is a kind of truthiness, since one can always trade off speed with the duration over which inconsequential effects are integrated. Some of these features are also discussed in Appendix 1. It turns out that the omission of any consideration of the Doppler in radiation pressure models is a major omission, but not necessarily for solar-sail modeling.

The physics of radiometry and of radiation pressure underpin how a radiation flux is measured and characterized. Specifically, a flux of incident photons is characterized by the spectrum of the incident photons and by the integrated energy per unit time and per unit area of the incident flux. Photons also have momentum, and when a photon is absorbed or scattered, the object upon which the photon is incident experiences a momentum transfer from the photon. For most non-atomic sized objects, the recoil from the incident photons is inconsequential. But, for solar or radiation sails, the mass per unit area of proposed sails plus their payloads is sufficiently small that the incident flux supplies a force that over time causes measurable accelerations. The smaller the mass per unit area being struck by photons, the more rapidly the objects are accelerated. Even larger masses such as satellites experience the slow inexorable accelerations from solar radiation.

However, recoil of electrons in atomic particles or recoil of nuclei can be substantial when experiencing an incident radiation flux. Solar panels make use of the recoil of the electrons within semiconductor materials that occurs when a flux of photons is incident onto a solar cell. In this case the electrons gain energy that allows them to transition into the conduction band of the semiconductor. In atomic transitions, we seldom speak of physical movement, but there is some because there is recoil from either reflection or absorption. But, in addition to simply supplying recoil to electrons, the incident flux also supplies a gross or classical radiation pressure on the object that may be either absorbing or reflecting the incident photons.

Some of the incident flux on photocells goes into developing currents, via the photovoltaic effect, and, therefore, the reflectance, while specular, is not one hundred percent of the incident flux. If all the photons were absorbed, there would be no reflection and the cells would appear black. In this case, there is still recoil of the electrons from the incident flux, which would be a momentum transfer from the incident photons to the photocells. If, on the other hand, the photons are totally reflected, the photocells would experience twice the momentum transfer, doubling the recoil, because the momentum of the incident photons has been reversed and that reversal comes from the recoil on the photocells. However, in this case there is also no photovoltaic effects. On the other hand, solar or photocells are never either one hundred percent absorptive or reflective, so there is a net recoil as well as power generation.

The small solar-produced radiation pressures have cumulative effects. Objects in orbit are in dynamic equilibrium with respect to the objects about which they are orbiting, which means that the centrifugal force on the object is balanced by the attraction of the central object on the orbiting object, which is a centripetal force. Solar pressure is a steady pressure, slight as that pressure may be, against such objects, and over time the small but cumulative force nudges objects out of their static orbits or trajectories.

For solar sails, the radiation flux is essentially perpendicular to the direction of motion. Solar sails are typically oriented face-on with the sun to capture the maximum flux and, consequently, the pressure is a slow push on the sail directed radially away from the sun. By canting a sail, a longitudinal acceleration is also possible, and this is how solar sails spiral inward or outward in their orbits. In satellites, large solar panels supply a significant surface area and the orientation and orbital parameters of some satellites are significantly and dynamically impacted by the solar pressure on the panels. We won't discuss orbital mechanics, but the small radial pressure forces a circular orbit into an elliptical orbit.

Once an object attains a radial velocity component relative to the sun, there are two effects that come into play. One is the Doppler and the other is a more fluid-dynamic effect that shows how the amount of radiation that reaches a moving object per second depends on the object's velocity with respect to the source of radiation. The fluid dynamic effect is like the effect of the wind on a balloon. When a helium-filled balloon is held by a string, a breeze will cause a pressure we can see and feel as the balloon is pushed by the wind. When the balloon is released, it is quickly accelerated to the speed of the wind, at which point there is no more force accelerating the balloon. This effect works both ways, in that motion into the wind would cause the effective pressure to increase because more air molecules are striking the balloon per unit time than when the balloon is moving along with the wind.

The Doppler occurs because the actual photon frequency when it strikes an object depends on the speed of the object relative to the photon flux. If we use a radar example, when an aircraft is moving away from a radar and is painted by that radar, the reflected return signal detected at the radar is Doppler shifted toward lower frequencies or red shifted. This shift is a function of the relative radial speed of the object away from the radar.

However, the photons have also given momentum to the aircraft, though it is too small to measure. Per Einstein, when the photons hit the object, the photon energy and momenta that are measured at the aircraft are already Doppler shifted because the aircraft is moving relative to the radar or radiation source. When the Doppler-shifted photon is reflected, the Doppler shift has already occurred because of the relative motion, and the momentum transferred to the object in the form of radiation pressure is determined by the Doppler-shifted photon frequency. The physics is that the actual interaction of the incident photon occurs in the moving frame. An aircraft is so massive that the aircraft rebound to incident radar pulses is negligible and the Doppler-shifted pulse's frequency is the same as the incident Doppler-shifted frequency. Compared to a single photon, all objects other than atomic particles are massive and we can ignore the change in frequency from the rebounding mass. This is not so when photons hit particles, but that is another analysis altogether, which will be discussed later.

Pressure is force per unit area and force produces a change in momentum, p. Thus, when photons of energy $E = pc$ strike a surface, we have that the photon momentum, which is E/c, is exchanged with the object being struck. For total absorption, the transferred momentum is just the momentum of the photon. If the photons reflect off the surface, however, the object supplies twice the magnitude of the momentum in the reflection direction because of conservation of momentum. Therefore, the totally reflected photons have momentum $2p$ opposite the direction of the incident photons, if the reflecting surface is normal to the incident radiation. However, the incident momentum is the Doppler-shifted momentum. That is, if the incident photon has energy $h\nu$, then the incident momentum on the moving object is $h\nu'/c$, where ν' is the Doppler-shifted incident frequency, as recognized by Einstein and as discussed in Appendix 1.

We get to the classical radiation pressure from flux by noting that flux is energy per unit area per second, or power, passing by some fixed point. In radiometry, we measure the power as watts per unit area from the sun at, for instance, Earth's orbit or on the ground. Given power, we find the total energy per second striking a unit and normal surface as P/c, where P is the power and c is the speed of light. (Normal means the direction perpendicular to the surface.) Since the photons' moments are given as E/c, we have that the radiation pressure is directly proportional to the incident power and, more precisely, radiation pressure equals power divided by the speed of light. Also, pressure is force per unit area, so the total force acting on a normal unit area is power divided by c, the speed of light.

Since flux is often measured at some point called a reference, we then find the power incident on a surface at some other distance r from the reference distance as a simple ratio of the range squared: $P_r = P_o (r_o/r)^2$, where subscript o is the reference value and r is the new radial range away from the source of radiation. The range in this case is not the range from the reference location. For instance, if the Earth's orbit is the reference range from the source, the new range is also taken from the radiation source. All natural flux diverges with range as $1/r^2$ relative to the source of that radiation. At great enough distance, all artificial sources also diverge as $1/r^2$ but the range r is different than for a natural source, such as the sun, and depends on the optics of the radiation source.

The Doppler and Radiation Pressure

We can call the form for pressure given above the static pressure model, and the static model is all that is currently used in radiometry or in calculating radiation pressure effects on space objects, even though a dynamic model was described within Einstein's 1905 paper. It is inexplicable that all the people who claim to have read that 1905 paper never noticed the dynamic radiation pressure model, which is an analogous circumstance to no one observing the issue with respect to Newtonian gravity. Such omissions occur, and they have consequences.

Now, let's see what the dynamic factors supply as modifications to the classic static model. In what follows, we speak not of the velocity directly but of the factor β, which is the ratio of the speed of an object divided by the speed of light: $\beta = v/c$. First, everyone has completely overlooked the relative speed term in the measurement of radiation. This factor is given as $(1 - \beta)$, and the relative speed term was derived from $(c - v)/c$, the ratio of the change in effective flux that can impact a moving object per unit time. We can refer to this as a compression factor. The above formula is for an object moving in the same direction as the flux. This can be generalized for any motion relative to the flux propagation direction as $(1 - \beta \cos\varphi)$, where φ is the angle between the direction of flux and the direction of motion. Thus, the effective flux decreases when an object is moving in the same direction as the flux but increases when the direction of motion is into the flux.

We can pause here for a moment to look at the effect of an emitting object moving relative to a receiver. In the last chapter, we showed that pulse arrival times are shifted by this motion. In the case of a steady flux such as from a luminous object, we ask whether flux from the relative motion of such objects changes with the direction of motion independent of any Doppler shifts? We reason out the answer in the following analysis.

If we look at two photons emitted in the same direction but Δt apart and from the same location on the moving emitter, these two photons are received at a time $\beta \Delta t \cos\theta$ apart, because the travel distance for the second photon is different than that for the first photon and where $\cos\theta$ defines the relative motion with respect to the direction toward a receiver. Therefore, in one second, a stationary receiver receives a different number of photons per second than if the emitter were not moving. Consequently, we have a compression factor that indicates that the moving emitter also has a compression factor comparable to that for a moving receiver. This change in flux is accompanied by the Doppler that is caused by the relative motion of the emitter. We discuss the ramifications of these two issues in Appendix 4 but otherwise ignore this factor in this chapter.

Einstein's Doppler expression from his 1905 paper, incorporating the same relative motion of an object through the flux as discussed above, is given as $v' = v(1 - \beta \cos\varphi)/(1 - \beta^2)^{1/2}$, where v' is the Doppler-shifted frequency of the initial frequency v. If an object is moving in the direction of the flux and $\varphi = 0$, we have the more familiar Doppler formula $v' = v[(1 - \beta)/(1 + \beta)]^{1/2}$. The Doppler causes the incident energy and, therefore, the incident momentum to be different from the static case, and the difference can either increase or decrease the incident power depending on the direction of relative motion.

For most velocities of practical interest, $\beta \ll 1$, and we can use certain approximations that are essentially good enough for all practical purposes. Using the small-β approximation we find that the first-order Doppler correction for the speed for the classical radiation pressure model becomes $(1 - \beta \cos\varphi)$. For solar sails, $\varphi \sim 0$, and the correction factor becomes $(1 - \beta)$. However, the first order correction for the compression factor is the same. Thus, for small-β, the impact of these two factors is $\sim (1 - 2\beta)$, which doubles the effect of motion on the radiation pressure.

Putting the above elements together yields a generalized dynamic radiation pressure model: $P = P_o (r_o/r)^2 (1 - \beta \cos\varphi)^2/(1 - \beta^2)^{1/2}$. For radiation sails, $\varphi \sim 0$, and the pressure model reduces to $P = P_o (r_o/r)^2 (1 - \beta)^{3/2}/(1 + \beta)^{1/2}$, where β is the speed of an object-relative to the photon trajectory. Later we show another application where this model changes our interpretation of other physics in a fundamental way. These ideas will be revisited in Appendix 1. Also, we will not make any use of the emitter compression factor, since the value of β for the emitter is not the magnitude of β used at the radiation sails and we often simply do not know the actual velocity of an emitter relative to the receiver. On the other hand, within the solar system, we can see that radiometric corrections might occur that are on the order of 0.0002 for various differential orbital motions of objects. Whether this is a measurable or important correction depends on the scenarios and the ability to make various dynamic measurements. Otherwise, we ignore the compression factors for emitters.

Going back to how Einstein derived his Doppler model, we can see that the physics would have been much different had he viewed the model from a photon flux perspective, yet that is how we make a measurement. Einstein continually uses the reference to an observer seeing something without regard to how they would see or measure anything. The fundamental element in his Doppler and aberration models is that he used waves and stated that they are still waves in all moving frames. He also stated that the physics of aberration was motion itself.

The Doppler and Radiation Pressure

We will look at the description of the Doppler in more detail later. But, the compression factor of a moving source causes both a Doppler shift as well as a flux change. The compression factors and Doppler shifts are identical in the term $(1 - \beta \cos\varphi)$. Therefore, the measured power can be as much as or less than twice that from just considering the Doppler shift alone. Later we describe how we can separate these two factors in measurements and show that we can distinguish between Doppler shifts and compression factors in the measured power.

However, the basis of being able to know that the Doppler or aberration exists is to make a measurement and measurements require radiation to interact in a detector of some sort. In this regard, a flux of photons does interact with the detector and that is where the Maxwell wave equation would be valid. Also, the aberration physics, discussed in some detail in the next chapter, are manifest in the solid materials with which incident radiation interacts, whether the materials are telescope mirrors or the detector. The incident radiation interacts with electrons within a material and that is also where Einstein's aberration arises. There is no "wave" as such in a vacuum, and the wave description appears to be a mathematical convenience and a mnemonic, since these equations do allow fluxes of photons to be quantified mathematically.

Now, the fact that Einstein did not complete or thoroughly investigate his radiation pressure model has had consequences. If we look at the dynamic part of the pressure model, which is plotted as a function of β in Fig. 4.1 when using the complete expression for the Doppler shift, we see that the driving pressure drops to zero as the speed of an object approaches the speed of light. We can approximate the impact of the dynamic terms for $\beta < 0.1$ by using $1 - 2\beta$, which shows how quickly the dynamic terms will limit the radiation pressure. What this means is that any object being accelerated by electromagnetic radiation can never reach the speed of light and its acceleration is far less than realized at speeds near c.

If we look at the dynamic radiation pressure model in more detail, we can see that for any sized object, since pressure equals force per unit area, we can show that the force accelerating a mass is proportional to the dynamic radiation pressure factors, which, as the object is accelerated to higher speeds, becomes smaller and smaller until the speed becomes balanced against any forces that may be slowing the object down. In other words, there is a terminal speed that is less

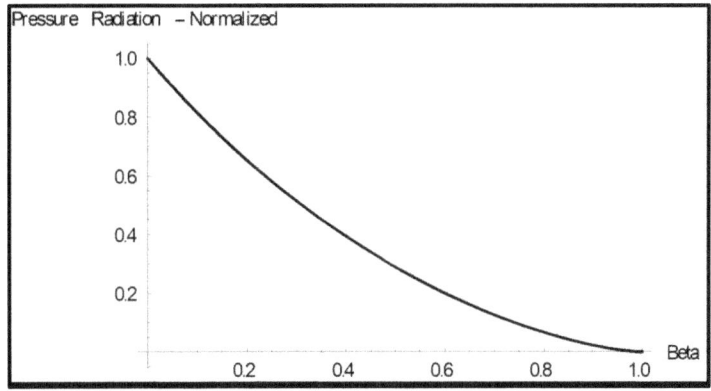

Figure 4.1: Normalized Radiation Pressure as a Function of Beta
(Normalized to zero relative velocity)

than the speed of light, which conditions are only met within particle accelerators or within some exotic astrophysical events. However, we now know that it is a fundamental limit within electromagnetics that defines the terminal velocities. There is no mass increase with speed nor need there be to explain observations.

For completeness, it is useful to see how Einstein reached his conclusion that mass increases with speed. Einstein's model for the kinetic energy in relativity was based on accelerating an electron very slowly in an electric field. An accelerated electron radiates energy and Einstein did not want to have to consider losses, so he made the acceleration so slow that there were no radiation losses. He found that the work done in moving a mass m to speed v was given as $KE = mc^2(-1 + 1/\sqrt{1-\beta^2})$, where the value $-mc^2$ is the constant of integration found for KE when speed equals zero and there is no kinetic energy. For very slow speeds relative to c, KE ~ 0.5 m v². From this we could infer that mass increases as $1/\sqrt{1-\beta^2}$. This form does not show a rest mass energy. The actual kinetic energy is still 0.5 m v², only now, per Einstein, $m = m_0/\sqrt{1-\beta^2}$, where m_0 is some mass when but the object has no speed. Thus, a rest mass m_0 is inferred but not an equivalent rest-mass energy as found in Minkowski's approach.

As work is done to move the object, Einstein showed that some of that work increases the speed and some supposedly increases the mass. None the less, he postulated a constant force and was calculating the kinetic energy at some velocity and assumed that the work done was equivalent to the kinetic energy. Consequently, we have the appearance that mass increases with speed.

The Doppler and Radiation Pressure

One of the classical experimental results motivating Einstein was a finding that accelerated electrons did not seem to be accelerated as much as expected for a given potential. The reason for the discrepancy was not known but was initially attributed to the mass of the electrons increasing as the speed of the electrons increased. Consequently, Einstein's expression for kinetic energy was taken as an immediate indicator of the success of his relativity, though we know from the radiation pressure model that what is really happening is that the efficiency of the process of accelerating the electron using the electromagnetic force is decreasing as the speed increases. It is a case of a correct observation with an incorrect conclusion. The fin-de-siècle physicists had no idea that force was the result of a flux, even for a fixed accelerating potential, and, therefore, it was not recognized that the decrease in the accelerating force was indicative of a radiation flux for which a Doppler effect could occur. More precisely, until Einstein's 1905 paper, we did not even have a correct electromagnetic Doppler expression and, having overlooked the radiation pressure model in the 1905 paper, the fin-de-siècle physicists never realized that they had another effect that might impact the acceleration of electrons moving at relativistic speeds.

From the dynamic radiation pressure model, we have that the force is dependent on the velocity, which we use for finding the work done to accelerate a mass via radiation pressure. If we assume that we are slowly accelerating a charge, just as Einstein assumed, to first order we can ignore radiation from the accelerated electron, just as Einstein assumed. Then, for constant radiation flux, assuming no range dependence, we know that the pressure is force per unit area. Therefore, we can multiply the radiation pressure both by some effective area for the electron and by the mass of the electron…or the mass of any object being accelerated…to find F = m a, which for electromagnetic acceleration contains all the terms in the dynamic radiation model. We are making the hypothesis here that a static electric field when measured is made up of a flux of photons. It is accepted that photons carry the force. Stated another way, we only know a static electric field exists when we measure it using an electric charge. If there is a force on a test charge, it is because of a photon flux, which implies radiation pressure.

The term *a* is the acceleration and contains some constant, K, and the function $f_1 = (1 - \beta)^{3/2}/(1 + \beta)^{1/2}$, which shows how the acceleration is impacted by the velocity. K is the constant accelerating radiation pressure (no range dependence) times the effective cross-sectional area of the object being accelerated. We can also re-write the expression for f_1 to cast it in a form that has some similarity to Einstein's expression: $f_1 = (1-\beta)^2 / \sqrt{1-\beta^2}$. From elementary calculus, we have that $a = dv/dt = v\, dv/dx$ and from elementary physics we have that incremental work, dW, is defined as F dx. Putting these together, we have that $dW = F\, dx = m\, K\, f_1\, v\, dv$, the increment of work needed to move the mass through a distance dx or to accelerate the mass by an increment of velocity dv. When we convert v to the variable β, just as Einstein, we have that $dW = m\, c^2\, K\, f_1\, \beta\, d\beta$, which we integrate from an initial velocity β_i to some final velocity β_f. The radiation pressure model also contains $m\, c^2$ because of the conversion to the variable β. The work integral is analytic but complex and does not yield a simple expression as with Einstein's model. For small-β, the results reduce to the standard form for kinetic energy. We can normalize the function for W using the value at $\beta = 1$, and the normalized function is plotted in Fig. 4.2.

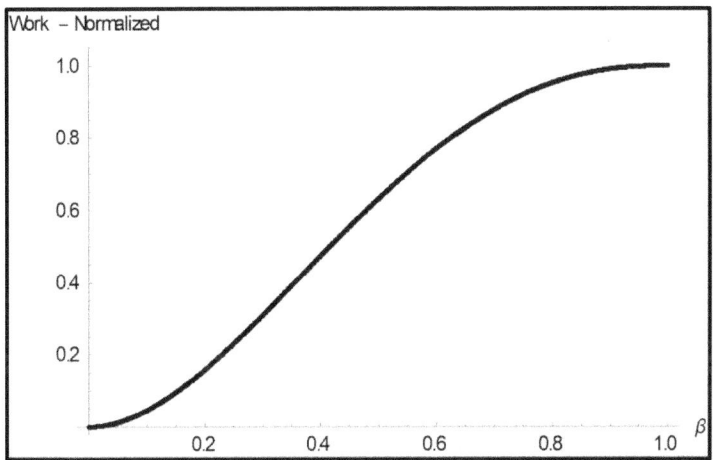

Figure 4.2: Normalized work done by a photon flux in moving a mass from β equals zero to one.

The radiation pressure model shows that the work stays finite, whereas Einstein's model showed that the work goes to infinity as the velocity approaches c. If we look at the definition of work, there must be an acceleration which shows up as a force and as a steadily increasing velocity. The radiation pressure model shows that the force goes to zero as v approaches c. Hence, the work to move an object from one velocity to another is always finite. In addition, the work to move an object from rest to c is also finite, which makes sense because the force function always stays finite.

It is likely, therefore, that Einstein's model is more related to the power required to attain a velocity, which would also be true with the dynamic radiation pressure model, since the amount of radiated power required to accelerate an object that is moving close to the speed of light becomes very large because the force becomes very small. Consequently, we need much higher powers operating over longer periods of time to increase the speed by some additional incremental value when we are already moving close to the speed of light.

The conclusion is that Einstein's model is not a model of real physics. The dynamic radiation pressure model only generally follows the same curve as Einstein's model in supporting an interpretation that the mass may increase with speed. We can also see that if we were to group terms such that we have m times f_1, it may appear that the mass is not constant but is a function of the speed. But, since f_1 is completely defined by the radiation model, there is no compelling reason to think that the mass is a function of speed.

The radiation pressure model also supplies an unrecognized factor that further limits the speeds that can be attained in particle accelerators or in outer space, which is discussed in detail in Appendix 1. However, the impact of this effect on particle accelerators is much more significant. The effect is a consequence of the local thermal ambient radiation incident on an accelerating particle. This is a totally ignored source of a force on an accelerated particle that acts to limit the speeds that can be attained.

Ambient radiation is the thermal blackbody radiation from the local environment or from the integrated radiometric flux from stellar objects. When all sources of ambient flux are added together, the ambient radiation exists all around an object as well as both in front and behind the object, where the rear hemispherical radiation can supply extra acceleration but where the radiation from the front hemisphere causes a resistive force. In the following discussion, however, we only consider the ambient flux as coming from directly ahead and behind a moving object, which means that we are "underestimating" the consequences of the ambient radiation in finding the radiation forces on an accelerating object.

The ambient radiation that is incident on the moving mass will undergo a Doppler shift. If we know the temperature in an environment, then for a particle or object moving through this environment, at any speed an object will experience differential forces from behind and in front, though as will be discussed later, there is a flux on all sides of the moving object and not just from directly in front or from behind that object. The radiation incident on the front of the particle experiences upshifted or blue shifted frequency changes, in which the frequency is higher than for a stationary object. Likewise, the radiation incident on the rear of the object is down shifted or redshifted to lower frequencies. The two compression factors are also different. The consequence is that the constant ambient radiation produces a net differential radiation pressure on a moving object that slows the object.

In interstellar space, and as more fully discussed in Appendix 1, there is an ambient universal microwave background as well as discrete stellar radiation from the stars all around the moving object. Thus, we can calculate a net radiation pressure by considering the direct stellar radiation flux plus the ambient thermal flux. While the stellar radiation is directional, it can appear to be quasi-random in origin far from stellar populations producing the flux and can appear to be like some ambient thermal radiation from stellar-temperature objects. The consequence is that the integrated stellar flux from discrete objects can have the same impact as the ambient thermal radiation, though the spectrum of this ambient radiation is given by some mean stellar flux, which is discussed in Appendices 1 and 4.

In addition, there is no range dependence on the thermal radiation flux, which is the same everywhere…which would also be approximately true for the ambient interstellar radiation far from any stellar object. At some interstellar distances, perhaps as close to the sun as within the Oort Belt, the local direct stellar radiation on an object would begin to approximate a uniform distribution from the forward and rearward directions, though this distance depends on the brightness of the local stars versus the integrated brightness of the distant general stellar population. This is also discussed in Appendix 1.

The ambient thermal radiation is called blackbody radiation and all the radiation frequencies in the spectrum are up or down Doppler shifted the same. We discuss this radiation in more detail in Appendix 4 in which we discuss the dynamic rather than the static blackbody radiation and how it should be quantified for moving emitters and receivers. At this point, the radiation analysis becomes more complicated, so we defer the details to Appendix 4. This is because the well-known Planck radiation model will be shown to be a heuristic and the way we calculate how the Doppler affects blackbody radiation appears to be wrong. Consequently, we continue the discussion in this chapter using first-order estimates of what we expect may happen and

The Doppler and Radiation Pressure

will restrict the discussion to quantifying how a flux is Doppler shifted and not how a distribution of wavelength in the blackbody radiation is Doppler shifted. We are only interested in the total integrated flux. Also, we have a fortuitous energy balance for certain scenarios that simplify the analysis. In the case of ambient stellar radiation, we have an estimated flux, so the Doppler shift on this radiation is easily found using the analyses discussed in Appendices 1 and 4.

However, since all incident frequencies are Doppler shifted by the same amount, the shape of a blackbody radiation curve does not change, so that the effect of the Doppler is not to create an effective change in the apparent temperature of the source but to change the effective distance away from the source of the blackbody radiation, which means that the effective flux changes in magnitude but not in spectral content.

To first order, we know the total radiated power from a blackbody at any temperature from the well-known model for the thermal radiation emitted from a unit area of a stationary object, which is $P_a = \sigma T_a^4$, where the subscript a stands for ambient, T is the absolute or Kelvin temperature of an object, and σ is the Stefan-Boltzmann constant. This radiated power from the unit area of an object is the same as the incident radiation on the object immersed within a thermalized environment at some temperature, such as within a kiln. At thermal equilibrium, an object absorbs as much radiated power as it emits.

In interstellar space, since all stellar objects have an effective color temperature, we would have that the distant and integrated radiation from all the stellar objects could be approximated by some effective color temperature. Depending on the distribution of distant objects at any given observation point, we would be able to estimate what amounts to an evenly distributed ambient flux onto a moving object. The closer stellar objects might supply an additional incident flux that is higher than the estimated "ambient" flux. Thus, to first order we can find the flux on a moving object that is incident from all directions. From the known motion of the object, we calculate the Doppler-shifted differential flux from which we can estimate the apparent radiation pressure slowing a moving object. This is also discussed in more detail in Appendix 1.

We have two scenarios where the differential thermal pressure may have an impact. The first is in space far from direct stellar radiation when some cosmic particle is moving close to the speed of light. We discuss this scenario in Appendix 1. The second scenario is in a particle accelerator, in which the particle beam is confined within an evacuated tube. These tubes are cooled to remove stray gas molecules. However, these tubes are not cryogenically cooled. We will look at the consequences of this scenario for an ice-cold tube structure and a cryogenically cooled tube structure with ambient temperatures of T_a=273K and 4.2K, respectively.

In most modern accelerators, microwave sources are accelerating particles and the ambient flux supplies a retro-force slowing the particles. We can, for analysis purposes, consider the driving radiation to be supplied by a constant flux. There are also other loses in the system that are measurable and would also retard the particle acceleration, but we will only look at the consequences of the radiation pressure.

Some analysis was performed using the Stanford Linear Accelerator (SLAC) parameters from which we can estimate what the driving microwave flux may be from the 250 microwave klystron sources. When we scale the klystron power to a waveguide tube cross sectional area, estimated for a 6 x 3cm waveguide, we get a flux of ~ 2.7Megawatts/unit area, which is essentially constant along the entire acceleration path, and we are ignoring what are called coupling efficiencies. Consequently, there are no range dependencies on the driving radiation pressure, only speed dependencies. We also ignore radiation emitted from the acceleration of the charged particles.

Note that the above estimates are likely quite far from the actual values. While the waveguide size is approximately correct, the actual beam tube is a hollow tube of some similar cross-sectional area, which allows the microwaves to be more efficiently coupled into and propagated down the beam tube, which has a size dependent on the wavelength of the microwaves. Hence, the above power/unit area is only approximate, but the chosen values will suffice to prove the point discussed in the following paragraphs.

To continue, the total pressure becomes P(Drive)+net P(Thermal), where the net P(Thermal) is negative. We need to remember that the driving power and ambient radiation fluxes are being affected by the Doppler and compression factors, so that the power delivered to the relativistic particles is much reduced from the static power. From one perspective, this is an analog to the differential power described previously. Setting the driving power $P_d = 2.7 \times 10^6$ watts/unit area and the ambient power as given by $P_a = \sigma T_a^4$ and applying the appropriate Doppler and compression factors to these two powers and equating them, we can find the terminal velocity to first order, neglecting any other sources of power loss. The terminal speed would be independent of the mass or cross-sectional area of the particle, which cancel from each side of the equality and which only comes into play when actual accelerations and velocity profiles are calculated. Any other system losses would simply reduce the terminal speed further. The terminal speed should also be independent of any other physical aspect of the particles, though in Appendix 3 we discuss some issues that make the above discussion only approximately true and only first order.

Equating these two fluxes, the drive flux and the ambient flux, neither of which has a range dependence, we find that the terminal values are $\beta \sim 0.979$ for $T_a = 273$ K (freezing point of water) and $\beta \sim 0.999995$ for $T_a \sim 4.2$ K (boiling temperature of liquid helium). It is likely that the current accelerators are not reaching their declared accelerations and particle speeds by significant values.

However, there is an issue with the above analysis. Once we approach relativistic speeds, the Doppler shift on the incident ambient flux can push the wavelength of this radiation well into the x-ray region. The experiments in upshifting the frequency of microwaves into the x-ray region using the scattering of the microwave radiation from electrons means that the ambient thermal background, which consists of wavelengths many orders shorter than microwave wavelengths, will be upshifted in frequency by many orders of magnitude higher than for the microwaves. When this happens, the scattering physics begins to change, so that the apparent cross section of the object may begin to experience deviations between those appropriate to microwaves and to x-rays. We look at some of these changes in Appendix 3, but it may be that the analysis above further underestimates the degree of slowing of the particles.

When radiation is incident on matter, we have macro and micro events that we must relate to the exact physics involved. Radiation incident on particles is essentially all scattered. When the photon energy is extremely large, though, an absorbed photon shatters the absorbing particle, and these effects and physics are not discussed or included in this book.

When radiation is incident on organized matter that consists of many particles or molecules, we can have scattering, absorption, and transmission of a flux of radiation. The radiation pressure depends on what is either absorbed or scattered. We usually measure these parameters as functions of the incident wavelengths of radiation. Thus, radiation pressure is a macro effect. The micro effect is associated with each scattering event for each photon. These types of considerations will likely not be an issue for typical speeds of matter in the solar system but can become an issue within particle accelerators and will impact what these devices are really achieving in the way of accelerations of particles.

In certain particle accelerators, the ambient radiation could be upshifted sufficiently in energy to cause the scattering physics to change. To first order, the driving microwave radiation's scattering cross section differs little from the upshifted ambient radiation's scattering cross section, though this is only approximately true for upshifted frequencies that are not significantly less than the Compton frequency, which is discussed in Appendix 3. For SLAC, we never get such large upshifts in frequency, which we can show using the above terminal velocities in the Doppler upshifted model. Consequently, the terminal speed calculated above does not need to consider different scattering cross sections and assumes that the scattering cross section is the same for both the driver and ambient radiation. This discussion is only included here so that these refinements are recognized as necessary in a detailed scientific analysis of the impact of the ambient radiation for certain particle accelerators.

In addition, the aberration, which is discussed in the next chapter, can cause the scattering angles to change significantly, so that the recoil forces are much more difficult to calculate. Considerable analysis and experimentation have occurred in quantifying what is called inverse-Compton scattering, in which a beam of microwaves is directed onto an approaching electron flux. This geometry is similar to radiation scattered from the forward location of a moving object, which is all Doppler upshifted and scattered in all directions. Such upshifting from electrons, in which the photons gain energy as the scattering particles (electrons) lose energy, has never been specifically applied to the ambient flux in the accelerator tubes. However, microwave energy beamed toward a flux of relativistic particles can produce a scattered and directed beam of x-rays.

For some reason the connection has not been made between this flux of microwaves and the presence of ambient blackbody radiation in the accelerator. The phenomenon of inverse-Compton scattering has been well documented and the concomitant slowing of the incident relativistic particles is also well understood, but, apparently, particle physicists have had tunnel vision in relating an incident microwave flux with an incident ambient radiation flux. Unfortunately, inverse-Compton scattering models use relativistic kinematic effects, which we have shown do not exist. Therefore, these models will need to be reworked to be applicable to quantifying the retardation forces within a particle accelerator.

The Doppler and Radiation Pressure

Since we are discussing particle accelerators here, we need to include some discussion of the aberration. The inverse-Compton scattering of microwaves from an electron flux experiences a beaming effect. What this means is that the scattered upshifted microwaves form an x-ray beam rather than a general scattered flux of x-rays. This occurs because of the aberration effect predicted by Einstein. When radiation is scattered from a moving object, the scattered radiation experiences an additional deflection into the direction of motion of an object. If the speed of the object is sufficiently high, a beaming effect occurs as the scatter radiation is given a pronounced forward momentum, which is an additional momentum transfer from the object to the scattered radiation. This means that the scattered x-rays supply a pronounced rearward recoil on the objects, which for particles increases the retardation forces on the object. Consequently, the way ambient thermal flux is scattered from the particles will have a pronounced recoil effect, further slowing the particles. From this we can deduce that the terminal velocity described above is much less than estimated. The conclusion is that the performance of particle accelerators is even poorer than estimated above.

If an optical spectral sensor is viewing the interior of the accelerator tube at the location in which relativistic velocities are being achieved, depending on the location along the tube, weak pulses of emitted actinic-like upshifted radiation would be detected that is synchronized with the pulses of microwave radiation driving the particle acceleration. Thus, we have a mechanism for detecting the passage of charged particles from the scattered and upshifted ambient radiation. The spectrum of the radiation would be an indicator of the true speed of the passing particles.

As a final analysis here on the Doppler, we will review how we measure a Doppler shift. In developing the dynamic radiation pressure model, we introduced a factor called the compression factor that is missing from all radiation pressure models. But the compression factor is only required if we are making a flux measurement. It is worthwhile exploring whether the lack of a measurement model is in fact a weakness in the relativistic electromagnetic theory. To test this concern, we will describe how to make a Doppler measurement.

Assume that we have a monochromatic flux and that we can perform some measurement to find the Doppler expression, such as making a measurement using a moving spectrometer. The moving spectrometer would display a line whose position and intensity varies as the relative speed is changed. While we have both the Doppler and compression occurring during the measurement, we can physically differentiate the Doppler change from the intensity change. Since the compression factor only changes the integrated intensity and not the Doppler, then, despite my mantra of needing a measurement model, it appears that, in principle, Einstein's approach to deriving the Doppler expression does not require a compression factor and is adequate as is.

What happens is that we find the scattering by putting our frame of reference on the moving scattering object, so that the object is stationary in that moving frame. We then use Einstein's Doppler model to show how the incident photons have increased or decreased in frequency just as they are about to impact the object. If the frequency shift is lower than for a stationary object, the interaction physics stays the same, which we show in Appendix 3.

The point is that Einstein's Doppler may be valid no matter how the photons and matter interact. There is, therefore, no reason to doubt the validity of Einstein's Doppler expression for any speeds. We also do not have to be shy about developing our models in a stationary frame, but we do need to be warry of simply applying a boost factor, defined later, to describe the post-scattered photons. It is more complicated than that.

Brillouin commented that Einstein's approach to the Doppler neglected recoil, which we have explicitly discussed in the radiation pressure models. On the other hand, the Doppler shift itself is separate from the consequences of the Doppler shift, including how we measure the Doppler shift and how we model the consequences of the actual interaction of a flux of Doppler-shifted photons with matter. The way we model radiation pressure, for instance, considers recoil, so Brillouin's concerns are not relevant. Mossbauer discovered an experimental way of accounting for recoil both during emission of photons and during absorption or scattering of photons All in all, we can proceed as we have in Chapter 4 and Appendix 1 and use the Doppler expression just as we have been as we discuss the impact of a flux of radiation on moving objects.

We could continue to dissect Einstein's 1905 paper, but that is not necessary. Most if not all the so-called relativistic effects can be traced to Doppler effects plus a lack of understanding of the physics of the experiments giving the "non-Newtonian" results. A choice was collectively made by the physics community that in retrospect was not a good choice, which was to view relativity as a kinematic rather than an electromagnetic theory. Even armed with the electromagnetic results from the 1905 paper, none of the "suspicious" experiments were ever re-vetted to determine if, after the revelations from the 1905 paper, the results were consistent. Nobody really read the 1905 paper to understand it. Thus, relativistic physics has stagnated at its pre-1910 state.

For one hundred years, the outrageous results of the 1905 paper plus its evolutions as supplied by Minkowski have been essentially accepted, embraced, and evangelized. With respect to Minkowski's contribution, it played to the mythology among mathematicians and theoretical physicists that the mathematics were the physics, which gave rise to a quasi-mystical belief that the more "beautiful" a mathematical model is, the more likely it is to be correct. As the major players within physics of the era bought into the theory, the incentives to question the paper receded until now it is anathema to the career of any physicist who questions the validity of special relativity. In other words, academic physicists are systematically violating the very academic charter under which they exist.

The unwillingness to question certain physics is a prelude to a kind of watershed within physics, only physicists have chosen not to engage in the battle and to ignore the watershed. As noted by Frank and Gleiser, there is "(A) Crisis at the Edge of Physics" but, as with scotomas, most physicists simply don't see the crisis or even believe in it. The consequence is that over the past century, everyone was looking for flaws in relativity in the wrong places…or else they really did not want to see the flaws.

At this point, we have identified the core of what special relativity really is, which is the Doppler effect and the dynamic radiation pressure. For most purposes, we could stop here. But, as we look at how the reformulated modern relativity impacts modern academic physics and some newer space systems, we need every aspect of the physics described by Einstein in the electrodynamic part of the 1905 paper. Our analysis of the radiation pressure has omitted a small "correction" factor. This small correction, alluded to previously, is called the aberration and is yet another mostly unrecognized source of perturbations on observed radiation pressure effects. We explore the ramifications of the aberration on radiation pressure in more detail in the next chapter and in Appendix 1.

What we show and discuss in the next chapter is that the source of the aberration has an observable impact on the dynamics of objects under the influences of external radiation fluxes. These influences are essentially perturbations on the results found using the Doppler expressions. None the less, these perturbations are observable and, in one case, may solve a century-old puzzle in solar spectroscopy. And, significantly, the aberration may represent a factor that can explain certain systems issues in optical devices. We show that the aberration may explain why observations that should be identical are, in fact, not the same.

Chapter 6—The Doppler Plus the Aberration

This brings us to the pièce de résistance in relativity, which is the Doppler plus the aberration model, which **is** Einstein's special relativity. The physics is subtle, but that subtleness has hidden much physics for a hundred years that would have addressed many observational issues that have remained unexplained. However, because of the subtlety and details, large portions of the material have been moved to Appendix 2. There are also many interesting scenarios and issues that will simply have to be deferred to some other books or papers.

It is inexplicable to me how the most famous paper in modern physics contains unknown and ignored physics that was overshadowed by the extravagant kinematics of the first relativity and within Minkowski's reformulation. For a paper that is so revered by physicists, Einstein's 1905 paper has been totally ignored and misunderstood ever since Minkowski reformulated special relativity. If the Kinematic Part of that paper is the foundation of modern relativity, then how could the rest of that 1905 paper be dismissed with little or no comment? Revering part of a paper and being totally ignorant of other content in that same paper may be a behavior pattern unique to academic physics. On the other hand, Minkowski's reformulation of special relativity essentially relegated electrodynamics to a single vector term that was, in essence, an add-on that never attracted the scrutiny it deserved.

In the earlier chapters, we explicitly took on the task of disassembling the kinematical theory. In Chapter 4 we showed that the electromagnetic portion of the 1905 paper was misunderstood, and I treated the undiscovered and unexploited physics as legitimate physics until proven wrong, since my efforts have only disproved the kinematical part. From a perspective of internal consistency, we can see that the electromagnetics represent sound physics as far as it goes in supporting the basic concept of special relativity. In Chapter 5 we applied what we had found in Chapter 4 to problems in applied physics and some technologies. These efforts were based on the notion that there was some there, there as described in Chapters 4 and 5.

As for the Doppler and aberration, support for Einstein's aberration is weak though we do have an aberration that is well known but, as it turns out, not well understood. The only direct validation for either the Doppler or the aberration has occurred for slow speeds that are a fraction of the speed of light, except for the inverse-Compton effect as measured in particle accelerators. The point is that the physics was there to be pushed and should have been identified and used by physicists over the past century. In pushing the physics still harder, I identify at least one definitive test for the aberration though not for relativistic speeds. For that we must rely on particle accelerators and the inverse-Compton scattering.

My approach to this chapter was to simply use certain elements of the electromagnetic physics in the 1905 paper as essentially correct as far as that physics went. In this chapter, I show how again we were misinterpreting what was in the paper and that we were essentially using the physics wrong. My mantra has been that we needed to reformulate Einstein's models into a form that supported measurement models. It was at this point, though, that I hit a speed bump.

Now, we need to discuss the aberration in some detail, since the aberration is a refinement or perturbation on the classical Doppler expression discussed in Chapter 4. This perturbation occurs because of the way photons do interact in matter and is not just a geometrical factor as is the astronomical angle-of-arrival aberration. When we make measurements, the photons are physically interacting with a measurement apparatus. In the case of radiation pressure, we have radiation incident onto a surface and being absorbed or reflected, and for all surfaces, the scattering is a combination of specular and diffuse scattering, which are a result of the interaction of the radiation with the electrons within the scattering object, as long as the photons are not energetic enough to knock the electrons out of their atoms. That is all we really must recognize.

In this chapter, we will be picking somewhat on the astronomers, because they have become the heirs to Newtonian gravity and relativity. In doing so, they have essentially adopted what physicists had accomplished without critically evaluating or improving upon these physics. As with technologist, they use what is given to them and have little interest or capacity to expand upon or critique these physics. If physicists do not recognize and improve upon the physics used by astronomers, then the physics will not get improved. In adapting and embracing general relativity, for instance, cosmologists essentially guaranteed that it would never be critically re-evaluated. I believe this is also true of the more mundane elements of relativity such as the Doppler and aberration.

The aberration is subtle and, to a certain extent, seemingly arcane. In fact, the aberration is essentially limited to academic observations, and there are few practical systems for which the existence of the aberration is of concern. On the other hand, there are some modern technologies for which the aberration may, in fact, be or become of some consequences.

One key point is that Einstein's aberration as contrasted with astronomer's aberration allows us to predict an atomic-level interaction between an incident electromagnetic flux with a moving physical surface, whether a mirror or a detector within some receiver. There is a physical underpinning to the aberration. However, astronomers have confused angle of arrival models, which describe one form of aberration, with Einstein's aberration model, and then they modified their simple models by using the linear ad hoc Lorentz boost factor $\gamma = 1/(1 - \beta^2)^{1/2}$ to make their model identical to Einstein's aberration model. In reviewing how astronomers arrive at their aberrations model, as with my initial review of relativity, the literature is a rat's nest of garbled physics…I stopped myself from declaring it gobbledygook, but a lot of it is.

We discuss the angle of arrival model below. However, a confusing fact is that Einstein's aberration is also to first order an angle of arrival model, but when we look more granularly into what happens when a flux of radiation is reflected from a moving surface or measured in a moving sensor, there is more physics involved than simply the deviation of the apparent image of a stellar source or any source of radiation. In other words, beyond the aberration model there is some physics that the aberration is demonstrating that we have essentially overlooked…yet again.

At this point it is useful to analyze the angle of arrival aberration model. The result of this analysis will supply better insight into how Einstein's aberration effect impacts how an optical system works and what happens to the Doppler within optical apparatus. The angle of arrival aberration derivation used by astronomers starts outside of Einstein's approach and is essentially baffling. To get the simple angle of arrival model to look like Einstein's model, astronomers add the "boost" or γ terms into the equations and invoke time dilation, addition of velocities, and even kinematic relativistic effects as well as referring to the motion of the source rather than the observer to justify incorporating these relativistic terms. Since we have identified these concepts as fictions, it is cleaner to simply use Einstein's approach, since every other approach results in the same models. But, to reinforce the notion that astronomers do not understand the aberration, they start with a classical angle of arrival model and then modify it to resemble Einstein's aberrations model. For small-β, all of these approaches reduce to the same form.

In a simple explanation for what we mean by angle of arrival, I will use an example that I at first took to be clever, then foolish, then cleverly simple, but finally concluded that it was explaining the wrong physics. Here's how the derivation proceeds. If we assume no relative motion and a telescope is viewing a stellar object, we can easily simply put the stellar image into the center of the field of view. If the telescope starts moving laterally, however, from the time the light enters the telescope to when that same light reaches down the length of the telescope tube to the mirror, the mirror has moved off laterally by a distance δ given by $\delta = v\,t = v\,d/c$, where $t = d/c$ is the time it takes for the light to travel down a tube of length d. Therefore, we can say that $\delta \sim \beta\,d$.

Now, since the image spot has been displaced, it appears that the light arrived from another angle. However, since the light is moving at an angle within the telescope, it travels a longer path than just d, the length of the tube, so the actual time to reach the moving mirror is given by $\beta\,d\,\sec\theta$. After some trigonometry, we arrive at $\tan\theta' \sim -\beta$ for small-β, where θ' is the new observed angle of arrival of the radiation. This is the first-order classical aberration angle which can also be derived from Einstein's aberration angle.

We can visualize the displacement of the image as the focal plane translating in the direction of motion and leaving the light rays behind. Therefore, the new image on the focal plane is displaced opposite the direction of motion. Consequently, from an imaging perspective, the interpretation is that the radiation has entered the telescope from an angle displaced toward the direction of motion. If we repoint the telescope slightly toward this apparent forward angle, the image will still be displaced but now it can be re-centered in the field of view.

However, for extremely narrow fields of view, the small tilting of the telescope can cause the telescope tube to block some of the incident radiation and an effect called vignetting can occur. If the speed of the telescope also increases, the aberration angle can also become so large that, even with repointing of the telescope toward the source, the aberrated radiation could miss the focal plane…in which case we have no information on how to repoint the telescope. The issue is that if the speed of an orbiting telescope is sufficiently large, the image from a star could be tracked until it falls off the focal plane. Granted, these are extreme conditions, but as we increase the apertures and narrow the fields of view of giant telescopes, these geometrical optical effects may become important, if they are not already.

There is a Wikipedia article on stellar or light aberration that takes the next step in identifying how the complete aberration model developed by Einstein can be deduced from classical physics using linear moving frames. However, astronomers then use addition of velocities or other relativistic effects to arrive at a "relativistic" model. I am hard pressed to understand why they took the step to invoke relativistic concepts to arrive at another expression that is the Einstein model. The rationale was likely that aberration is relativistic, and it would not do to have a classical-looking derivation that simply had to be wrong…somehow. In a way, by modifying the angle of arrival model to match Einstein's model for the aberration, there seems to be a denial that Einstein's physics are correct or at least relevant. Why modify the classical model to be the model they wish to obtain? The issue seems to be that from a teaching perspective, Einstein's work is too complex for "beginners", so some further baffling manipulations were used to get to Einstein's model.

The simpler first-order classical model developed above is wrong from the perspective of how to derive it and does not show how radiation scattered from a non-optical surface such as a radiation sail or spacecraft is aberrated or even how the aberration occurs in something like a radar or microwave antenna. The classical model emulating Einstein's model is $\tan\theta' = \sin\theta/(\cos\theta - \beta)$, where θ' is the deviated or aberrated observed angle and θ is the actual arrival angle of the radiation if $\beta = 0$. These models will require further discussion later as we identify how θ and θ' relate to the angles φ and φ' used by Einstein in his aberration model, which we discuss below.

I am only taking the discussion of the aberration far enough to identify how to proceed both in mending our models and in interpreting observed phenomena. There is simply too much history…a century's worth of efforts… that will need to be revisited to determine if or how the aberration might be important. I start by "explaining" what the aberration really is and how to use it, but then I relegate specific examples to Appendix 1 and simply describe some things that need to be examined in more detail and why. Otherwise, I leave it to future efforts to identify whether these suggestions represent new opportunities, are wrong, or are irrelevant.

Before getting explicitly to the aberration, Einstein's derivation for the aberration and the Doppler shift may have been the subject of misinterpretation for a hundred years. The error is not in how the models were derived but in how they have been used. Again, the misuse was simply not important from a practical perspective. Consequently, yet again, we had no reason to suspect something was amiss, though there were and are suggestions. Now we need to look at what the Doppler and aberration expressions are telling us. It turns out that we have been sloppy in using these concepts and mathematics.

For reference purposes, Einstein's **Doppler was given previously as** $v' = v(1 - \beta \cos\varphi)/(1 - \beta^2)^{1/2}$ and the aberration from the 1905 paper is given as $\cos\varphi' = (\cos\varphi - \beta)/(1 - \beta \cos\varphi)$, where the angles φ and φ' are not obvious and we need to look at how they impact observations. There is a distinction between these two angles and the actual angles used in making observations, which we discuss a bit further on. We need to be clear on what these angles are and what they signify.

In addition, while I use the term Doppler for various models, such as the model for v' above, that usage is also somewhat sloppy. I raised this point earlier but repeat it here so that we are clear on what is being modeled. The Doppler is the change in the measured frequency because of relative motion. Einstein's models merely allow a new frequency to be found when measurements are made from a moving platform. The Doppler shift is the difference between the measured stationary and moving frequencies. In the above model, the Doppler shift or simply the Doppler is $\Delta v = v' - v$, which in the small-β approximations is $\Delta v = -v\beta\cos\varphi + v\beta^2$, where we have only kept terms in β and β^2. The term $\beta \cos\varphi$ is simply the first-order projected velocity along the line of sight and β^2 is the second-order relativistic correction which is a result of the aberration, which we get to shortly. However, in Appendix 2 we show that the aberration has a much larger impact on the Doppler shift than the first-order equation above for Δv indicates.

We show the angles pictorially in Fig. 5.1, only one set of which is explicitly defined by Einstein as the angle between the direction of propagation of the radiations and the direction of motion of the observer. If we are on Earth and looking at a stellar object, we are in the moving frame, not the emitting object's frame. Einstein's derivation was all about an incident wave or flux of radiation and a comparison between what a moving observer and a stationary observer measure at the same point in space and time. Since the primed values are what a moving observer measures, then we measure v' and φ'.

Einstein's Relativity

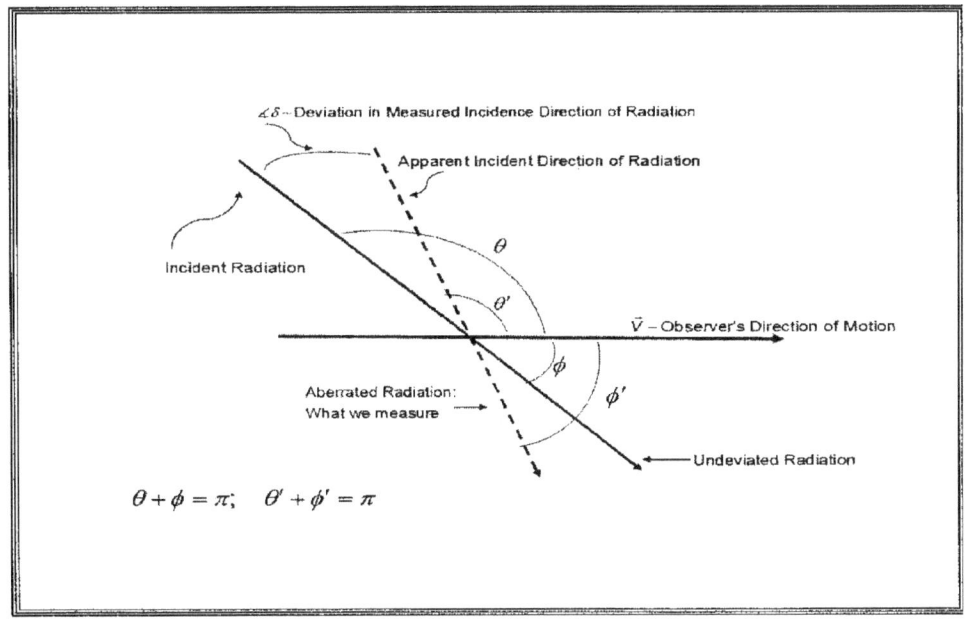

Figure 5.1: Definition of the Angles used to define the aberration.

We use Fig. 5.1 to show how Einstein's defined angles compare to the angles used in making a measurement. They are different angles but are related to one another. The main difference is that a moving observer must be looking toward a radiation source to make a measurement. As usual, this was a practical element that simply did not enter Einstein's modeling. However, the key point is that Einstein's model contains the angles φ and φ', whereas from a practical perspective, in making an observation, we would use the more common angles θ and θ', as shown in Fig. 5.1.

Figure 5.1 shows how the various angles compare in Einstein's approach to developing his Doppler and aberration models. We need to relate the two sets of angles and see if there is any impact on the models from our using one set of angles rather than the other set. An astute reader will see that Einstein's Doppler expression is given in terms of $\cos\varphi$, whereas we need to insert the expression for $\cos\varphi'$, because we are moving through the radiation and measure φ'. We can find the correct expression by solving the aberration expression above for $\cos\varphi$ as a function of $\cos\varphi'$. When we measure spectra, we measure ν' and from that would be inferring ν…if we even bother. Later we investigate whether, in fact, it does make any difference whether we think we have measured ν or ν'. And, when we measure φ' we infer φ, which is, of course what the astronomically observed aberration is all about. But as will be pointed out, we never truly concern ourselves with the angles φ and φ' in a practical sense, since our data are compared with calibration data. At the end of the day, however, we must anchor these relative data to the absolute data. This anchoring is when it matters which model is used and why.

Consider the way in which Einstein derived the Doppler and aberration models. Figure 5.1 provides a description of the geometry Einstein used. The direction of propagation is not the direction from the observer toward the source of the radiation being measured. Since the observer needs to be looking toward the source of radiation, we define a new angle θ' called the look angle, where θ' is the look angle toward the deviated radiation, which we measure, since we are in a moving frame. The un-deviated radiation is coming from an angle θ. Note that no particular care is given by astronomers to whether they are measuring θ or θ' when calculating the Doppler shift in a measured spectrum, but it does matter despite many astronomers making no distinctions even when measuring cosmic spectra, though even in this case we will show that we are essentially using a small-β for the aberration despite the measured Doppler shifts being associated with tremendous receding velocities. Keep in mind that for all practical purposes, the aberration is usually associated with perturbations on the direct radial Doppler shifts.

Recall that Einstein derived his models for the Doppler and aberration without considering how an actual measurement would be taken. Therefore, we need to transform the Doppler model to be representative of how we make the measurement, which is to view things at the angle θ'. This means that we must transform Einstein's Doppler expression given in terms of φ into the expression given in terms of θ'. We then compare the results of using this correct expression with that given in terms of φ where we are simply and blindly plugging in the value of θ' for φ.

At this point I will insert an observation from further along in this chapter and something that is remarked upon in Appendix 1. The aberration is noted as a deviation in the arrival angle of radiation toward the direction of motion. This is what Fig. 5.1 indicates occurs. In the simple classical aberration model relating to the time for radiation to propagate down a moving telescope tube, the stellar image falls behind the movement of the focal plane. Therefore, the apparent angle of arrival is from in front of the same image obtained using a stationary telescope. At this point we have no interaction of the radiation with any physical material such as a detector at the focal plane. Consequently, the only physics discussed so far relies on simple geometry.

There is a purpose behind this discussion. It has proven difficult to ferret out exactly how astronomers arrived at the Doppler model that they use. It appears that, in effect, they ignore how the Doppler expression is derived and simply use the look angle θ, ignoring the fact that we should at a minimum be using angle θ' in place of φ. However, we compare the model used by astronomers with the correct model to see if it makes any difference what angle or model we are using. I used the book by Kitchin in understanding how astronomers generally use the Doppler and perform spectroscopic measurements, though this book is not readily or cheaply available.

We can make the angle conversions to find the Doppler shift in terms of the look angle θ', which is given as:

$$\nu_c' = \nu(1-\beta^2)^{1/2}/(1-\beta\cos\theta'),$$

which can be compared with Einstein's original Doppler expression given previously, which is:

$$\nu' = \nu(1-\beta\cos\varphi)/(1-\beta^2)^{1/2}.$$

The correct model is the one given in terms of θ', since that is the look angle we actually measure. If we take Einstein's Doppler expression as is and replace φ with θ, where $\varphi = \pi - \theta$, we have $\nu' = \nu(1+\beta\cos\theta)/(1-\beta^2)^{1/2}$, which is the model used by optical astronomers. It is not clear whether astronomers recognize the complete role of the aberration, but they do apparently realize that the angle φ used by Einstein is not the look angle θ. Furthermore, there is no recognition that it is θ' and not θ that is what we are actually measuring. It turns out that the angle difference between θ and θ' is essentially inconsequential to spectroscopy with a single exception. Otherwise, to first order, these forms predict the same Doppler shift. It is at the second order where the forms differ. But, in those cases in which β is not small, we would need to be careful that we are using the correct model, since the second-order terms are no longer inconsequential.

Now, after the above discussions and criticisms, there is another issue. We have no difficulty understanding that the Doppler shift is a result of the relative velocity of an object with respect to a source of radiation. We further recognize that the incident radiation into a sensor is arriving essentially perpendicular to the input aperture of the device. The aberration, on the other hand, if it is truly a result of the interaction of the incident flux onto the surface of the object, depends on the velocity component of the sensor's velocity that lies parallel to the surface of the mirror or optical element within the sensor. This is the perpendicular or transverse component of the velocity of the sensor with respect to the direction toward the source of radiation, or $v \sin\theta'$. Here we are making the case that Einstein's statement that the aberration is caused just by motion is too simplistic to mean anything. The aberration must also have a physical origin.

In the discussions on radiation pressure, we noted the orientation of a surface relative to the incident direction of the radiation was important. Usually we look at the projection of a unit area toward a radiation source, so that the incident angle may not be, and usually is not, perpendicular to the surface. However, the aberration is essentially independent of this angle in that all radiation incident on a surface is aberrated the same. The Doppler explicitly identifies v $v \cos\theta$ as the primary Doppler that occurs. In the small-β approximation, we can show that the aberration angle is $\delta \sim \beta \sin\theta$, which is the component of the velocity of the sensor or an object perpendicular to the flux, and, therefore, the component of the relative velocity can be considered to roughly be along the surface of some object intercepting the incident flux.

Consequently, we can make the case that the aberration is a result of the interaction of the object with the transverse component of incident flux. This is consistent with the physics of optics. Since reflection or absorption requires an interaction of radiation with the electrons within the material, from conservation of momentum, the photons reflecting at some angle relative to the surface normal requires a rebound of the object, since the reflected photon has been given a new lateral momentum component. Therefore, Einstein's aberration predicts some interaction and some new photon momentum with a concomitance loss of momentum by the scattering surface or object.

Let's look at the aberration models in a little more detail. We start with the so-called classical derivation of the complete aberration, which for small-β reduces the deviation in an image to $\delta \sim \beta \sin\theta'$. When we convert Einstein's aberration model from angles φ and φ' into a comparable expression in θ and θ' and then find $\tan\theta'$, we can compare Einstein's aberration to the classically derived aberration model given previously, which is $\tan\theta'(classic) = \sin\theta/(\cos\theta - \beta)$. By finding $\tan\varphi'$ from $\cos\varphi' = (\cos\varphi - \beta)/(1 - \beta \cos\varphi)$, using the identity $\sin^2\varphi' + \cos^2\varphi' = 1$, we find that

$\tan\theta'$ (Einstein) = $\sin\theta/\gamma(\cos\theta + \beta)$, where γ is the ad hoc term that makes the classic derivation match the relativistic form, only these two derivations differ in the sign of β. Note that β is referenced here to the direction of the flux of radiation. The mismatch of terms in the models further confirms that astronomers do not understand the physics of the aberration. Since this physics has been essentially given over to the astronomy community, there are no physicists who are concerned with the correctness of these models, and the astronomers simply use them without knowing what they truly represent.

Getting back to the Doppler expression, we still need to see how astronomers measure Doppler shifts. The process of measuring a Doppler is via comparisons of measured spectra with calibration spectra, and often the calibration spectra are generated in real time within the spectrometer itself with calibration lamps, or at least this used to be the process. These technologies and techniques are very well developed and expertly used, but as we will discuss, there is more going on than realized within spectrometers.

Spectrometers contain complex optics plus a disbursing element such as a grating, which physically separates radiation spatially by wavelength or energy. We know the amount of disbursing as a function of wavelength, so that a Doppler shift is also a shift in the measured wavelength. Consequently, if we can measure physically on film or on the now-ubiquitous CCD images some shift of a wavelength from its calibration spectra, then by using our models we can determine the relative Doppler velocity component. This is where the models come into play, which is in the final determination of the relative radial velocity component. In practice, use of the calibration lines is well documented, but the comparisons have historically been a matter of tedious physical measurements to quantify the image or spectral displacement. Ultimately, everything boils down to the quality and accuracy of the calibration spectra and the use of correct models to identify what the spectra mean in terms of relative velocities.

However, a review of the literature did not indicate that there was any awareness that the angle astronomers call θ was the wrong angle to use or that some further transformations were required to cast Einstein's Doppler model into a practical measurement model. The incorrect Doppler that was found is $v_i' = v(1 + \beta\cos\theta)/(1 - \beta^2)^{1/2}$, where we have simply replaced Einstein's angle φ by θ. For small-β, which is almost the only values of β that astronomers need to consider, we have two versions of the Doppler, which are:

The Doppler Plus the Aberration

$$\nu_c' \sim \nu (1 + \beta \cos\theta' - \beta^2/2) \text{ (correct)}$$

and

$$\nu_i' \sim \nu (1 + \beta \cos\theta + \beta^2/2) \text{ (astronomer)}.$$

We ignored terms in powers higher than β^2 for now, which only comes into play for very high relativistic speeds. Despite its seeming irrelevance to spectroscopy, we needed to use the aberration equation to find the correct Doppler model based on how we make measurements. In fact, the final forms for the various Doppler expressions take on the forms they do because of the aberration. Otherwise, the models would only consist of the first-order term $\beta \cos\theta$. The higher-order powers are only important for certain specific cases, most of which are of academic and not of practical importance…though this may be changing as discussed toward the end of Appendix 2.

What the correct Doppler model shows is that the frequency has a small permanent red shift that is only dominant when θ' approaches 90 deg. The astronomers' model, on the other hand, has a permanent blueshift at all angles. The difference between the two models approaches a maximum of β^2 as the look angle approaches 90 deg. Therefore, the relativistic aberration factors only become important for certain look angles that are essentially perpendicular to the direction of motion of the Earth or the observer relative to the incident radiation, if these small values can even be measured. But the situation is more complicated than this. Later we show that there are hidden sources of perturbation in our measurements and interpretations that could significantly impact what we think we are measuring.

We now have everything about the Doppler and aberration that is needed to determine when and how the aberration is important. In most practical instances, it is not significant. It does have an impact, though, but only for certain academic sciences and technologies. In Appendix 1 we show that the aberration does supplies a measurable long-term radiation-pressure perturbation on satellites and spacecraft. In addition, there is a yet unknown and unquantified impact of the aberration on particle accelerators because of the large β within accelerators, which has an impact for further reducing the terminal speeds of particles in these devices. There may also be an impact in certain high-precision optical systems used in defense and commercial applications, especially if there are resonant cavities for which there are multiple reflections of radiation, such as in laser or microwave devices. We will comment on a few of these below and in Appendix 2.

We are limited in how we can measure things and then in how accurately those measurements are. Sometimes the things we hope to observe are simply buried in the signal because the overall physics requires more subtlety than our technologies can provide. Consequently, in the following discussions, while there may be mathematical differences between the models and how we use them, we may not be able to measure these distinctions. Even so, we need to be aware of the distinctions because they again play to the theme that it is not wise to ignore physics just because at some point it may seemingly be inconsequential. It is still missed physics, and that fact is forgotten over time…if it was ever recognized.

So, addressing the distinction between the above models, we can find the frequency error between the two models, the correct and incorrect models, in terms of wavelength by noting that since $c = \lambda \nu$, then $0 = \nu \Delta\lambda + \lambda \Delta\nu$. Consequently, if we know $\Delta\nu$ for some frequency of radiation, then $\Delta\nu/\nu = -\Delta\lambda/\lambda$. But, since $\lambda \nu = c$, then we have that $\Delta\nu = -c \Delta\lambda/\lambda^2$. In optical spectroscopy, we measure λ and $\Delta\lambda$ but in radio astronomy, we measure $\Delta\nu$ and ν. The Hubble Space Telescope, for instance, which has a wide range of optical spectrometers, can achieve ~ 0.01 nanometers resolution, depending on the instrument and wavelength region, which translates into the velocity range Δv ~20-100 km/sec for the Doppler resolution, which we show how to calculate in Appendix 2 and which depends on the wavelength. Over the UV to IR range of wavelength, we can have a range of a factor ~ 5 in the effective velocity resolution for the same wavelength resolution.

In contrast, ground-based systems for extra-solar planet exploration and for solar spectroscopy claim to be able to detect relative Doppler shifts down to ~ 0.1m/sec of motion. Therefore, while the imagery from the Hubble is extraordinary, there are some measurements that simply cannot yet be made from such space-borne systems. But, because of the quality of the exoatmospheric imagery from the Hubble and the lack of both atmospheric absorption and image distortion, there are measurements that can made in orbit that cannot be made on Earth.

It is useful to see what the above differences in the models means in terms of the error in estimating the speed of an object, since the Doppler does occur because of the relative speed difference between two objects. We are interested in the error from using the wrong model, so we have $\Delta\nu = \nu_c - \nu_i = -\nu \beta^2 \sin^2\theta'$,

The Doppler Plus the Aberration

where we have gone back to the exact forms, formed the difference, and then looked at the small-β approximation. The maximum error $\Delta\nu$ is on the same order as the aberration, which is given as $\nu\,\beta^2$. Because β^2 is usually so small, the impact of the relativistic terms is ignored. However, as pointed out in Appendix 1, this assumption is not always true. We also discuss later in this chapter a method of validating Einstein's aberration model through measurements in which we can measure the relativistic Doppler term.

Nevertheless, in high-resolution Doppler spectroscopy, we can resolve these small perturbations, so that measurements could be biased by the aberration effect. The above difference in the models would result in a minimum bias in the radial Doppler $\sim$ 2m/sec, which is only resolvable in high-resolution terrestrial spectrometers. However, the Doppler measurements in searching for extra-terrestrial planets are typically relative measurements, so the bias is not relevant to the accuracy of the interpretation of the results unless the look angle is changing as the spectrographic data are collected.

Recall, though, that we have only found the difference between using one or the other of the Doppler expressions to extract the velocity associated with a measured Doppler shift. Both expressions for the Doppler show that at $\theta \sim \theta' \sim 90$ deg we still have a non-zero Doppler shift. This shift is variously referred to as the transverse or zero-crossing angle Doppler (zero because there is no obvious component of radial velocity), which is nominally not zero, but as we will show, it in fact does go to zero at a specific angle. The transverse or zero-crossing Doppler is a result of the aberration.

In one reference, there was a statement that there is only one object bright enough at 90 deg angle to observe the aberrations impact on the Doppler. That hardly seems possible. However, the fact remains that this is the only angle for which it may be possible to observe what is called the transvers Doppler shift, which is discussed further below. The real issue is that when using the correct Doppler expression, the maximum correction is only - $\beta^2/2$, which may or may not be observable using some advanced specialized spectrometer. This being the case, there are few academic circumstances in which it matters that the angle being observed is θ' rather than simply θ. Except as discussed in Appendices 1 and 2 and later in this chapter, there can be a real impact from the aberration that has not been noted so far.

However, that there is only one object in the heavens that is observable at 90 deg is not precisely a correct observation. There are two orientations in which the radiation from a stellar object is perpendicular to the direction the Earth is moving. One location is perpendicular to the solar ecliptic. The other location is in the plane of the ecliptic in which the Earth's orbit is crossing the radiation from a star. We have, therefore, two orientations for which we can measure the transverse Doppler, where the orientation perpendicular to the ecliptic may, in fact, have only a few useable stellar objects at the proper location. However, there is a significant band of stellar objects that lie in the plane of the ecliptic and it is likely that some of these would provide opportunities to measure the transverse Doppler from a telescope with a high-resolution spectrometer that could detect the small Doppler at the critical crossing angle.

But, despite the apparent paucity of observable stellar objects at the 90 deg location, there is one stellar object for which we can achieve exactly 90 deg relative angle every day of the year, and that is the sun. Our solar telescopes have high enough resolution that we can easily scan the surface of the sun, such that a scan from the leading to trailing limb of the sun spans ~0.53 deg or ~ 9.25 milliradians. Large aperture telescopes have angular resolutions far smaller than the solar disk, which allows a spectroscopic scan of the sun to be highly localized and precise because of the brightness of certain spectral lines.

As an example, highly ionized iron has been studied for a century or more and has presented a dilemma to spectroscopists. As the solar disk is scanned limb to limb, the spectrum of solar iron shows a larger redshift in the spectrum at the limbs of the sun than at the center. There is no explanation for this issue, which is called the solar redshift problem. But by better understanding the aberration, possibly we can explain the anomalous observations.

If we look at the correct Doppler expression, we see that as θ' approaches 90 deg, the Doppler shift does not go to zero but to - $\beta^2/2$. The astronomer's model would show + $\beta^2/2$. This Doppler shift is called the transverse Doppler, because visually the radiation is arriving at the moving Earth at 90 deg relative to the orbital motion of the Earth. Attempts to make this measurement on Earth using Earth-based or laboratory sources have proven futile, because the physics is for the sensor to be moving perpendicularly through the incident radiation. This needs to be re-emphasized, because experiments are conducted in which the source moves or both the source and receiver move, but only when the receiver is moving can the aberration be detected.

More detailed investigation of the Doppler expression shows that when we set the Doppler to zero and solve the model, we find that there is an angle θ' for which there is no Doppler, and this angle is at $\beta/2$ relative to 90 deg. When viewing the sun, this angle identifies a true zero-velocity Doppler, since we have no Doppler shift at all. This angle has also proven impossible to measure and resolve in laboratory measurements but not when making direct solar observations. The prior discussions show that the observer or receiver must be moving fast enough such that the local value of β results in a resolvable angle for which the small Doppler shift can be resolved. This is possible using solar spectrometers.

A scan across the sun results in a minimum Doppler that is off-center from the observed center by the angle $\beta/2$, and scans in either direction across the sun from this point show that the spectrum of iron becomes red shifted in both directions of scan compared to the center of the sun. Thus, the spectroscopy of the sun supplies a test of the aberration and Doppler models. But there is more to it than this, which requires an additional discussion of the aberration.

We also need to quantify what the impact of the aberration is on measured spectra. Recall that once a photon flux has been reflected from a surface, such as a telescope mirror, it follows a straight path with no lateral inertial "drift" of the path associated with the lateral motion of the reflecting surface. The radiation is aberrated, which is an angle adjustment in the direction and path that the photon follows, but the path origin is "anchored" at the initial aberrating surface. The propagation of radiation through additional optical elements should also cause the same aberration effects at each surface, since the optics of lenses and mirrors are defined by the interaction of the photons with the electrons within the material. We discussed some aspects of this issue in Chapter 2, where the apparent lack of an interaction between radiation and transparent media was discussed. There are apparently differences between aberration from a lens and from a mirror, where we can count each mirror as a surface for which the aberration effect will occur. At some point, experimental data will need to be obtained to place a limit on the magnitude of any aberration caused by the motion of transparent crystalline media.

However, there is apparently no pronounced aberration from within transparent optical materials, so that the final detector is the only surface that seems to introduce aberration in a refractive telescope other than within a spectrometer that may have multiple reflecting surfaces. However, the evidence for how radiation interacts with transparent media is solid but not well understood. It is obvious that one hundred percent of the radiation incident on opaque or reflective materials interacts with the material. On the other hand, except for the reflection of radiation from transparent media, which is a result of the interaction of the radiation with the media, the mechanisms on how radiation interacts within transparent media is not well understood other than to be a result of quantum effects.

The purpose of this discussion is that the total system aberration is system dependent, in that multiple surfaces incrementally add to the aberration, which in turn changes or perturbs the measured frequency. Therefore, the cumulative aberration within a system also causes a change in the aberration-induced Doppler perturbations within a system. Thus, two seemingly similar systems might show different results from spectroscopic or aberration measurements, which in high-resolution system could be measured and introduce an unknown bias in the interpretation of the measured results. There is more discussion on this in Appendix 2.

At this point, the details are far more arcane than warranted for this book, though we do look in more detail on some other aspects of the aberration in Appendix 2. In other words, at this point the discussion becomes even more abstract and drowned in detail, if that is not already the case. The details are subtle but also help to explain why certain types of measurements have never been successful carried out…or have they? While this may seem cryptic, our laboratory measurements have failed in specific instances while our astronomical measurements have, in an unrecognized way, performed the same experiment. When correctly interpreted, the astronomical observations may help to solve, in one case, a hundred hear old mystery regarding the solar spectral redshift and help to resolve the issues regarding why particular measurement have simply proven impossible to perform in the laboratory and why seemingly identical instruments do not supply identical results.

As for practical consequences, the aberration only represents a "correction" or perturbation to the basic models, though it does predict some dynamics that are otherwise omitted. We discussed some of these effects in Chapter 5 and in Appendices 1 and 2, in which some radiation pressure effects on spacecraft can be found or interpreted by incorporating the aberration effect into the various models. Other instances in which the aberration may play a role is in explaining certain orbiting system orbital perturbations. These perturbations are a result of the aberration causing photon reflections to be different than we expect, with cumulative deviations that cause moving systems performance to differ from the static performance.

The Doppler Plus the Aberration

For instance, since the path of a photon or a flux of photons interacting with a surface is aberrated by any transverse motion, possibly a space borne laser or atomic clock would experience a different integrated magnitude of β than existed when the device was calibrated or aligned. Any aberration of the radiation reflecting from the laser cavity mirrors could cause the radiation to walk-off the mirrors and cause the actual beam direction and frequency to become perturbed compared to the same values measured on Earth. In fact, such consideration of the aberration's impact on laser performance could extend to airborne systems, in which the optical aberration in the pointing, tracking, and fire control system may inadvertently insert unexpected biases in certain advanced system.

The discussion in Chapter 2 on Cahill's and astronomical observation results shows that the Earth likely has a cosmic velocity > 600km/sec, which supplies a maximum β of ~ 0.002 to be factored into any perturbation of the spectra from cosmic objects. Clearly, any cosmic redshift would only be negligibly affected by the aberration, though as pointed out in Appendix 2, optical systems can multiply the magnitude of the aberration effects simply by the way the radiation is collected, imaged, and measured.

Something that has not been noted, however, is that Einstein's Doppler and aberration constitute a conservative system. By equating the complex phases for stationary and moving observers, we form a conservation equation. Therefore, the aberration can be considered a momentum conservation requirement, which is supported by the physical mechanism of the photons being scattered by the transverse motion, conserving he momentum of the moving object and the photon flux.

At this point, the essential character of the aberration has been presented along with the potential consequences of ignoring or being ignorant of it. From the discussions in this chapter, certain aspects of astronomy are more ad hoc and inaccurate than we are led to believe. The main thrust of this chapter is that we have yet another element of physics that has been ignored or overlooked. The fact that this physics comes from one of the most highly regarded papers ever written adds credence to the idea that there is a blight in modern physics. Yet my goal in this book was to get to the point at which the narrative will demark into technology, which is the subject of another book, *A Novel Propulsion System*.

Chapter 7—The General Theory of Relativity

All the information needed to show that there is no general relativity as currently understood has been presented, though not necessarily discussed completely. There were two fundamental hypotheses that were the underpinnings to special relativity and there are three that are fundamental to the underpinnings of general relativity. For special relativity, the two assumptions or hypotheses were that the speed of light is frame independent and that all laws…or the representations of these laws…are also frame independent. For general relativity, there are two additional hypotheses or assumptions. One is that the Minkowski four-vector was a true representation of space-time or four-space and the other is that the equivalence principle was valid.

We talk to the equivalence principle below, but here we can see that there is also an overlap in that the laws or representations of the laws are frame independent and that the speed of light is frame independent. Recall that a frame is any moving coordinate system. The hypothesis that the laws and their representations be frame independent takes on a more rigorous interpretation for general relativity. Yet it is not all that clear what the possibilities of a variable speed of light might mean. Add to this ambiguity the fact that the support for a variable speed of light is weak and contradictory. Consequently, we will maintain the postulate that the speed of light is constant everywhere.

The theoretical physicists during the fin-de-siècle period believed that all legitimate representations of the law of physics must be both covariant and Lorentz invariant. Lorentz invariance is what was applied in the Kinematic Part of the 1905 paper, but covariance is a tensor property for a much more general representation of the laws, specifically the space-time form of any representation. An irony is that any representation can be cast as either a covariant or contravariant tensor, and the two representations are related through transformation matrices that also obey certain rules. Consequently, the motivation for having all representation either covariant or contravariant was a weak motivation. None the less, it drove the reasoning of the theoretical physicists in looking for specific representations.

In fact, the seminal paper published by Einstein in 1916 on general relativity was essentially a tensor lecture. The topics in the paper started with introductory concepts and proceeded to the most advanced topic in general relativity, and most of the underpinnings to the paper were supplied by Einstein's friend, the mathematician Marcel Grossmann. While no solutions were offered for the equations that Einstein derived, it was indicative of the lack of general mathematical expertise within theoretical physics at that time that motivated Einstein to develop and publish a tutorial in order for the intended readership to actually understand what he was doing.

With the above brief discussion, we can move on and summarize the above by simply noting that everything about both special and general relativity was wrong. The Lorentz transforms do not work on anything but EM radiation. In addition, the Minkowski four-vector defining space-time as four dimensional was based on a mnemonic with no real foundations other than it allowed the Lorentz transforms to be derived. Once Minkowski had identified the space-time form for s^2, since this form was both Lorentz invariant as well as covariant, it was firmly established by the theoreticians of the time as a valid representation.

The discussion relating to the original kinematic relativity shows, however, that the four-vector as defined by Minkowski was a mathematical fabrication that seemed to fit the requirements for special relativity. It predicted things that were not part of the original 1905 paper. While that is not in and of itself an issue, the problem is that it also predicted things that were part of the kinematics of the 1905 paper. Since the prior discussion has shown that those elements of relativity were themselves false physics and based on misinterpretations, then anything duplicating those predictions must also be false physics.

Let's reinforce the above conclusions as to the validity of the Minkowski four-vector. Minkowski had found another way to interpret the equation $x^2 - c^2 t^2 = 0$. If we have a straight line defined in some rectilinear coordinate system, the axes of that coordinate system are given by the unit vector symbol, which for the x-axis is $\hat{x}$ and for the time axis is $\hat{t}$. If the line lies along the x-axis and is defined as s in length, the vector for this line would be $\vec{s} = x\hat{x} - ict\hat{t}$, where Minkowski called time the fourth dimension. The length squared of this line is then $s^2 = x^2 - c^2 t^2$, and he further noted that comparing the line in unprimed with the primed frames could be represented as $s^2 = s'^2$. This description of s^2 is Lorentz-transform invariant and is how Lorentz initially derived the Lorentz transforms.

By further calling the term $-ict\hat{t} = \vec{u}4$, we can write the general four-vector form of the line as $\vec{s} = \vec{u}1 + \vec{u}2 + \vec{u}3 + \vec{u}4$. Furthermore, the length squared of a vector is found by a vector multiplication process called the dot product identified as $\vec{s} \cdot \vec{s} = u1^2 + u2^2 + u3^2 + u4^2 = x^2 + y^2 + z^2 - c^2t^2$. Therefore, Minkowski created a new four-vector with time as the fourth component and he never had to determine where this form came from and why we needed to work with the square of a length and not the actual vector. This is yet another manipulation, since the non-squared form of the vector $\vec{s}$ had its origin in the EM propagation representation, but Minkowski's new four-vector and square of the new vector rendered the origin irrelevant to the physicists of the time. The net result is that propagation has all become identified with kinetic propagation representations., even the propagation of photons.

At this point, physicists were happy to work with the square of the length, since in this new four-space the magnitude of s^2 was only zero under special circumstances, which is the propagation of EM radiation. In addition, s^2 was at first glance the same mnemonic as used by Lorentz but with an added dimension, though the form still had no real origin other than it seemed to work. And that is the issue. Minkowski was simply manipulating forms that satisfied mathematically rigorous rules, and Poincaré had elevated this rigor to new levels that further elevated the status of Minkowski's efforts. Once the rules for establishing mathematical rigor were met and once the manipulations gave the same results…and more…as supplied by Einstein's 1905 paper, no one ever questioned the manipulations, which still occurs in academic theoretical work.

However, since we have shown that his form should originate with the complex phase of propagating radiation, we would expect a form more like $x - ict = 0$, where i stands for imaginary. (When we square i we get -1 as a value.) Therefore, squaring both sides we have $x^2 = -c^2t^2$ and $x^2 + c^2t^2 = 0$. We are unable to recover the original form for s^2. Even if we factor $s^2 = x^2 - c^2t^2$ to eliminate cross terms in ict, we have two different solutions which are $s^2 = (x - ict)(x + ict)$, which when multiplied together still does not recover the original expression. What we have, then, is simply mathematical manipulations focused on obtaining expressions relevant to special relativity. Since Minkowski's approach does duplicate the kinematic effects Einstein found, it cannot be physically correct, since there is no evidence that time dilation or spatial contraction actually exist other than as a result of manipulations of data.

All the manipulations were pursued because of the belief that the Lorentz transforms would work with mass as well and EM radiation. This belief drove everything, including the belief that mathematical expressions or representations had to adhere to certain mathematical rules. And, to further the perspective of mathematical rigor, Poincaré showed that the Lorentz transforms satisfied rules associated with what are called groups. The theoreticians of the time…and now…firmly believe that the mathematics and their rules define the allowed equations and, therefore, the physics.

Einstein also developed a postulate called equivalency. He speculated that we could not distinguish between freefall toward a gravitational source and freefall within a coasting spacecraft, in which we are comparing an inertial reference…the coasting rocket…and a non-inertial accelerated frame…the freefall in a gravitational field. This so-called equivalence is broken by velocity measurements, in which the velocity changes during freefall toward a planet but not for the rocket that is coasting. Consequently, contrary to Einstein's hypothesis and belief, we can distinguish between the two. In addition, Eötvös and others had shown that there was not a significant different between inertial mass and gravitational mass, which gave rise to what is called a weak equivalence. On the other hand, it is likely that equivalency was merely used to help Einstein focus on the ideas of geometry in defining gravity. It seems irrelevant that equivalency was an incorrect concept.

Einstein, using the ideas associated with the equivalence principle plus the Minkowski four-vector for space-time, developed his theory of general relativity by forming a generalized tensor description of Newtonian dynamics in four-space and by incorporating the acceleration from gravity with that of a kinetic description. The form for s^2 was identified with a metric called g_{ij} and was cast into a more generalized differential form as $ds^2 = g_{ij} dx^i dx^j$, which was a tensor relationship that will not be discussed any further other to recognize that the term g_{ij} is called the metric of the space, which defines the curvature of the geometrical space upon which the differential lengths dx^i and dx^j are defined. The differential lengths are identified with generalized coordinates in the space defined by g_{ij}.

Next, Einstein defined what is called the Einstein energy equation from the 1905 paper that defines kinetic energy as $E = m_0 + T$ and T is just the kinetic energy of a mass with mass defined as $m_0/\sqrt{1-v^2/c^2}$, where v is the velocity of the mass, c is the speed of light, and m_0 is called the rest mass. When E is combined with a potential, such as the gravitational potential given as U = - GmM/r, the Newtonian gravity potential, then the Lagrangian, L, is formed, which is L = E – U. The Lagrangian is used in what is called the Euler-Lagrange equation given in tensor (generalized) form to find the geodesic or lowest energy path of a mass moving in the space defined by the g_{ij}. Whew!!!

Paths that objects follow as they move are defined by the metric g_{ij} of the space or manifold in which the mass is constrained to move, and the paths are called geodesic paths. Consequently, Einstein was able to develop a Lorentz invariant equation for motion that duplicated the acceleration of gravity. In other words, the equivalence principle allowed Einstein to justify combining inertial behavior with accelerated (non-inertial) behavior as far as the motion of mass was concerned.

As an example of how we find geodesic paths, we can describe paths on the surface of the Earth using the coordinate system of the surface of the Earth, which is a specific metric g_{ij} that is defined only on the surface of a sphere. We can then find the shortest path between two points on the surface of the Earth, which is the geodesic path, by using the Euler-Lagrange equation, and these paths are known as the great circle routes or paths between the two points on the surface of the sphere. Solving the proper equations defines a disk with its center at the center of the Earth and with its radius the same as the radius of the Earth and oriented so that the two points on the surface of the Earth fall on the rim of the disk. The curved line lying on the surface of the Earth connecting the two points is a geodesic path and is the shortest path between the two points. In addition, the motion of an object along this shortest path also defines the lowest-energy and preferred path the object would follow.

Using the equivalence principle, Einstein generalized his geodesic equation to include the paths objects follow under the influence of gravity using the Minkowski metric for space-time. The equation used to find the geodesic curves was first derived in the mid-seventeen hundred's for finding the geodesics or minimum lengths of curves that objects follow between two points, including such things as the arc of a rope suspended from two points under the influence of gravity. The shape of the hanging rope is called a catenary and is the same curve seen in the cables of suspension bridges.

None the less, things are even more complicated. The interpretations of the Minkowski metric is to describe paths objects follow in four-space or in space-time. In other words, an object is propagating in space-time even if it is not physically moving, since time continues to "flow". The whole idea of the

Minkowski metric is to define the paths objects follow in four-space, which is space-time. No attempt is made to identify a representation for which we could define propagation as in the EM portion of the 1905 paper. What has occurred is that the Minkowski metric has been applied to support kinematic-like propagation for mass and EM radiation, and $s^2 = 0$ describes EM radiation propagation. In order to grasp the subtlety of the philosophical and mathematical efforts that went into inventing the equivalence principle, the Wikipedia article "Equivalence Principle" supplies an abundance of background for understanding Einstein's mindset.

If the above sounds confusing and confused, it is because general relativity was an attempt at justifying something that was unjustifiable and was not real and gave rise to what in now called pseudo-science. From the above description, we can see that general relativity requires special relativity to be true and to be described by the Minkowski four-vector or metric. It took the combined efforts of generations of theoretical physicists to weave the nearly seamless presentation of general relativity. The development was driven by the equivalence principle and by the results from special relativity that requires all laws or mathematical statements of physical activities to be frame independent, covariant, and Lorentz invariant. Once the equations were manipulated into supplying some results that seemed to explain observations as well as to predict something observable, the issue then became how to justify these ad hoc creations. It is from these discussions and justifications that most physicists were and are bullied into believing that there was or is some there, there.

More interesting is that even Einstein admitted that all the predictions of general relativity could be made using Newtonian gravity, though some magnitudes of effects would differ from the general relativistic predictions. Since we have no useful or unambiguous observation for general relativity… note that the operative term is unambiguous…and since the metric used to develop the equations is a fabrication, one would have to question why we keep training more physicists and cosmologists in general relativity and continue to support attempts at observations that cannot be made and likely can never positively support general relativity. General relativity has become an obsession and, in Gordin's words, immunized from deep scrutiny.

Brillouin also stated that Einstein's general relativity equations were so general that they only supplied a framework for modeling scenarios. The most common solutions to the equations are the spherically symmetric solutions first derived by Schwartzschield. Brillouin pointed out that there were other spherically symmetric solutions to Einstein's equations, though he did not discuss why these other solutions have been ignored. These other solutions yielded different radii for the event horizon, but since we have never actually measured an event horizon and the associated mass, any of the solutions would be correct in that regard, current observations included. We have seen "something" but only interpret what that something was in terms of an existing theory or hypothesis.

Additionally, there have always been competing theories for general relativity, but all start with the Lorentz metric. In the end, even these competing theories have disappeared in the past fifty years, since they offer nothing that is not consistent with Einstein's general relativity, even after 100 years of attempting to make definitive measurements. Clifford Will, a true believer, has summarized the historical development of general relativity and other competing theories. It was only in ~1960 that cosmic observations began to suggest that general relativity could have a role in explaining newly observed phenomena, such as quasars. On the other hand, Will does not go into the reasons that classical gravity would not have sufficed, since one can find an event horizon for a sufficiently massive classical mass, though this event horizon is twice the radius of that predicted by Einstein's theory. Since we have never seen an event horizon, or at least we have no direct measurements of the mass of an object purporting to have shown an event horizon, it is still unclear why Einstein's theory was necessary, except for supplying a putative explanation for the cosmic redshift.

There other aspects of the cosmos, such as the redshift and expansion of the visible universe, that are admirably address by general relativity. These features are not explained via alternate theories to the "big bang" origin to the universe, such as Fred Hoyle's steady-state theory. In the steady-state model, matter is continuously created throughout space, thereby pushing the existing galaxies apart and laying the groundwork for the creation of new galaxies to further populate the universe. Hoyle coined the term "big bang" to, as he commented, create a simple verbal method of contrasting the two existing theories to a radio audience. To me, the idea of continuous creation has the same legitimacy as positing dark matter to explain other gravitational effects observed throughout the cosmos, including the observation that the redshift seems to be increasing the farther away the observed objects reside. Therefore, this raises the issue as to whether the universal gravitational constant is truly constant?

In *Newton's Gravity*, the generalized Newtonian gravitational law was applied to finding the mutual force between a small test mass embedded within an extensive uniform distribution of mass. The results are non-intuitive and in some scenarios in stark contrast to the same results using the point-mass mutual force. Figure 7.1 shows the mutual force on a very small test mass embedded within a uniform spherical distribution of mass. The test mass is moved from the center of the sphere out to the outer edge, which could be considered an analogous way of looking at the force on a stellar object from all of the other mass within the visible universe. The generalized Newtonian gravity model is identified as a Machian model, which Einstein was never was able to incorporate within his general theory. It is odd that Mach's perspective was never recognized as identical to that of Newton's, which is the origin of the generalized Newtonian gravitational model. Newton's actual description for finding the mutual force between extended objects was overshadowed by Newton having resorted to the point-mass approach in calculating the mutual forces.

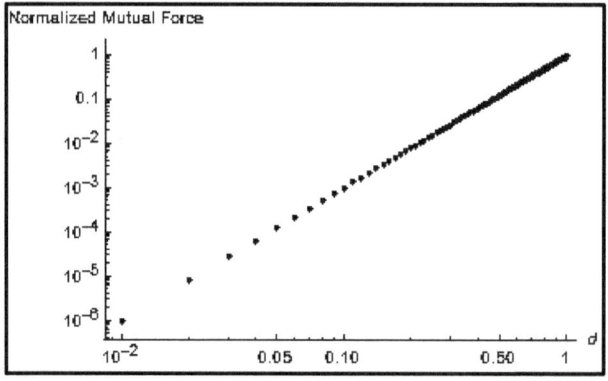

Figure 7.1 Normalized internal mutual force on a small test mass embedded within a uniform spherical mass distribution. Normalization is with respect to the point-mass model at each distance d from the center to the outer edge of the spherical distribution. The test-mass radius was 10^{-10} times the spherical distribution's radius, which was set at unity.

Figure 7.1 demonstrates that the generalized model is not very close to the point mass model for finding the internal mutual force, which contrasts with the variance in the external mutual force between uniform spheres when comparing the generalized model with the point-mass model. For non-spherical objects, such variances between the two models was often substantial when comparing the generalized model with the point-mass model, sometimes producing results completely unanticipated by the results of using the point-mass model.

The purpose of the above discussion is to note that the cosmic steady-state model, which now has few adherents, can be re-interpreted to incorporate some of the same observations as met by general relativity. We know that photons are affected by gravity, gaining and losing energy as the photons travel toward and away from massive objects, respectively. Therefore, the light from the sun would be redshifted the further from the sun an observation is made of the sun. For a vast cosmos, any observation point is equivalent to any other. At the limits of the visible universe from the Earth, the mutual force on the photon increases as the photons propagate to the far edge of the universe, thereby increasing the redshift the further the photons travel. Consequently, photons propagating toward the Earth from the far edge of the cosmos are likewise redshifted, since these photons would experience an increased mutual force the further that they travel from their sources.

Hoyle's notion that the creation of matter between the galaxies could be likened to injecting gas into a vacuum. Without some confining gravity, fluids cannot develop a pressure and will expand without limits. Consequently, the newly created mass would exert an outward flow of mass, thereby causing the cosmos to expand. Yet the expansion may be a false interpretation of the way the redshift increases as a result of the gravity experienced by photons as they travel outward toward the limits of the cosmos. We also have the supposition that mas does not shield mass with respect to gravitational attractions. We have no measurements that suggest that mass shields mass, though the redshift may be a manifestation of just that occurring over cosmic distance. The issue is that the redshift seems to be isotropic within the universe and that any object anywhere observes the same quantified degree of isotropic redshift. No classical…that is, non-relativistic…models exist that seem to meet observations unless there are other assumptions that are classically valid but are not considered possible.

The general form for Newton's gravitational law is unknown and has not been applied to the redshift issue, either with or without the hypothesis that mass does shield mass from the effects of gravity. No efforts were attempted in this book to quantify these assumptions other than to note that the general Newtonian gravitational model can be considered as another model that predicts behavior heretofore only quantifiable via general relativity.

The General Theory of Relativity

The conclusion reached earlier in this book is that there is no general relativity as we currently understand it. Einstein's approach is a fabrication based on other fabrications and misinterpretations. While the ideas of space-time associated with Einstein's general relativity are intriguing, the actual representation is pseudo-science at best. Yet there may be a better theory that maintains some of the persuasive elements of Einstein's theory. Brillouin pointed out that the notion that gravity waves will propagate at the speed of light is at best also arbitrary, though Einstein's theory addresses the thorny issue of gravity forces propagating instantaneously, which was something Newton questioned. Personally, while I find the concepts within general relativity to be interesting, I find the science of the current general relativity to be little more than pseudo-science. With a little effort, we can produce models that match observations to some degree that rely only on classical gravity.

Finally, other than the conundrum of the cosmic red shift, only one other observation within the solar system purports to support the concept of general relativity and that is the advance in the perihelion of the orbit of mercury. It is claimed that there is no classical explanation for the advance of the perihelium, which is too small to observe or measure for any other solar orbiting object. In reviewing the general equation with the classical equation using the general Newtonian mutual force, we find that both mathematical descriptions contain a small perturbation of the order of 10^{-10} in magnitude. In *Newton's Gravity*, we showed that the mutual force depends on the size, mass, and distance between two objects, so that general relativity, as a point-mass representation, does not truly represent the potential and forces on objects, whereas the generalized Newtonian representation does clearly show that both objects contribute to the interaction and that there is no such thing as a potential field associated with just one mass. In addition, the deviation of the sun from a sphere supplies another perturbation.

Consequently, there are unexplored aspects to the cause of the advance of the perihelion of Mercury's orbit that rules out this orbital perihelion advance as a demonstration of Einstein's general relativity. It has been understood for several centuries that any deviation in the mutual force representation away from the exact inverse-square of the distance between two interacting and orbiting objects causes the perihelion of the orbits to precess. Consequently, if the general Newtonian mutual force representation had been known, there would have been no reason to believe that Einstein's general relativity was the only explanation for the orbital perihelion precession.

Chapter 8—Einstein's Relativity: Some Conclusions

The most basic conclusion in wrapping up the analysis in this book is that Einstein's 1905 paper is both more and less than we have recognized. It was not accepted in its original form, but once Minkowski had re-formulated it, the new models for relativity were both more elegant on the one hand and less correct on the other hand than the same physics as described in the 1905 paper. The physics community was smitten by the mathematics and for a hundred years the quality of the resulting theories was never fully compared against those initially proposed in the 1905 paper. While the natural philosophers of the time were fully capable of critiquing the 1905 paper, none ever undertook the task of re-vetting that paper against Minkowski's new mathematics. Perhaps they could not, in that the mathematical skills of the physicists of the time were inadequate, which may be supported by Hilbert's observation that physicists did not know enough mathematics to be good physicists. Of course, Hilbert, mathematical savant that he was, also got it wrong.

There is occasional recognition that some elements of the 1905 paper were either wrong or irrelevant. For instance, Weinberg comments in the article "The Revolution That Didn't Happen" on the fact that we now know that Einstein's longitudinal and transverse masses, which Einstein defined in the 1905 paper as he derived his relativistic kinetic energy, are not actually real distinctions, and the only mass is rest mass. In researching the concepts of these distinct types of mass, I found that the distinctions are resolved in much the same way as astronomers rationalize their aberration models, which is by using convoluted descriptions of classical relativistic effects. Such rationalizations are still occurring. The work in this book shows that the entire modeling in the 1905 paper regarding kinetic energy was not the modeling of real physics.

What Weinberg seemed to be saying in the article is that ultimately it is irrelevant how the pioneers presented their new material in fomenting a paradigm change, since subsequently and by consensus, once the paradigm shifts, no one cares what the pioneer had said in the particular ways they said it. Newton is a good example in that nobody does physics the way Newton did physics. But what we showed in the book *Newton's Gravity* is that we should have paid closer attention to what Newton was saying and not what he was doing.

What Weinberg implies is that Minkowski's reformulation of relativity was the new paradigm and that Einstein's efforts were simply a consolidation of prior efforts by the likes of Lorentz and Poincaré. As it turns out, Minkowski is seen to have developed the first and most mature or modern part of the new paradigm that was finally fully defined by Einstein's general relativity. Once the paradigm shifts, no one looks back. Kuhn argued that they could not look back for various technical and social reasons, namely that, for instance in the case of relativity, they could no longer understand what Einstein had produced in the 1905 paper...which I believe is a sophomoric interpretation of what really happens. What really happens is that a sociology develops that inflicts professional pain onto those who violate the strictures of that sociology.

By contrast, Weinberg would argue that scientist can look back but that they seldom do, since there is no reason to do so. If the discussions in this book hold, then we do have two solid reasons to look back critically at the roots of paradigm shifts. Not wanting to look back is a social and cultural issue. Physicists are discouraged from looking for mistakes in those theories that have been accepted by the consensus and are now essentially accepted as the core of the new paradigm. Paradigms are not to be researched for flaws but only to discover ways forward toward the next paradigm. Paradigms are never withdrawn as would a bad publication with apologies all around.

It seems clear that the physicists during the fin-de-siècle period were grappling with new concepts and simply lacked a reference for more carefully vetting the 1905 paper. The ideas in simultaneity, for instance, are both trivial and a complete misinterpretation of what we now understand as elements of system modeling, queuing theory, and operations analysis, which are ways in which complex activities are modeled to identify logistical, logical, and information deficiencies. Simultaneity had its basis in an information-deficient set of scenarios as well as the erroneous applications of the Lorentz transforms. More practically stated, we can surmise that you can only know what you can know, and any lack of information creates ambiguities that are essentially unresolvable without other assumptions or information, which must later be shown to be correct based on outcomes of events. Academic physics has been intellectually sloppy over the past one-hundred years by disregarding an inevitable truth, which is that all branches of science should be continuously putting their older science and paradigms under a microscope, and in Kuhn's words, retesting the origins to a paradigm....no matter how painful and distasteful that may be.

More egregiously, the supporting experimental data at Einstein's disposal was misinterpreted on the one hand and wrong or misrepresented on the other. The concepts of relativity were simply beyond the understanding of the scientists of the time. The ultimate irony is that the 1905 paper contains within it what was needed to finally articulate and understand relativity. It is a case of a bootstrapping effort in using something to understand itself. Is that possible?

All the evidence for justifying the 1905 paper was based on a kinematical perspective even though Einstein was accidentally incorporating radiation propagation representations in trying to understand his kinematic models, and his electromagnetics were interpreted as kinematics. And to me a most glaring deficiency is not recognizing that the electromagnetic theory essentially leads one to understand how to derive the Lorentz transforms without resorting to the Lorentz mnemonic for the square of the length of a propagation path as viewed in two different moving frames of reference, to whit $r^2 - c^2 t^2 = r'^2 - c^2 t'^2$. The electromagnetic models show that we can use $x - ct = x' - ct'$ just as well in finding the Lorentz transforms, which would have essentially shut Minkowski's approach down and avoided a century's long misinterpretation of what, exactly, Einstein was driving at in the 1905 paper. A little thought shows that if the Lorentz transforms work to solve $x - ct = x' - ct'$, then we should be able to find the transforms from that same identity.

We also have the M-M experiment as being both misinterpreted and poorly executed. If Cahill's assessment of the M-M experiment is correct, we are led to conclude that Michelson and Morley were incompetent and deceptive in presenting their "evidence", perhaps self-deceptive, and were merely sycophants in search of respectability within a larger community to which the American physics establishment of the time were hardly competent to participate.

We also had an incorrect, or at least incomplete, understanding of what we mean by inertial references. We had arbitrary assumptions about what we could and could not measure with no real justifications. Such concepts as the equivalence principle were poorly conceived and, therefore, became a false litmus test for what we could and could not measure. By denying the analytical model for detecting motion within a moving frame, we arbitrarily established a belief that velocity could only be inferred from measured accelerations. For a century, we have ignored an accumulation of evidence that absolute motion could be measured, though, ironically, it is not absolute motion in the sense of being referenced to a hypothetical rest frame at the far ends of the universe. We missed the logical conclusion that our universal reference frame is still relative but is relative to a local reference, which is defined by the very photons we use in making measurements.

The broader conclusion is that, while academic physics has gotten certain things terribly wrong, there is also within academic physics a unique unwillingness and inability to re-visit the past. While it is human nature to resist being proven wrong or admitting to being wrong, academic physics has embraced a more corrosive perspective in refusing to acknowledge that they may not only be wrong but also that they are unwilling to even consider the possibilities. While the vetting of new work is often subjected to scathing criticisms, once the work has passed muster, especially the work of the grand poohbahs of academic physics, the physics is forever treated as infallibly correct. Gordin and Popper have commented on this tendency within academic science, and this tendency is the subject of the essay by Frank and Gleiser as they lay the blame for much of this attitude in physics at the doorstep of certain branches of theoretical physics. The mathematics is NOT the physics as most theoreticians believe or are led to believe.

So, we can now wrap the discussions up. The only relativity is that associated with electromagnetic radiation propagation. While we have applied the physics of radiation pressure in describing various scenarios, we can draw some other conclusions from Einstein's 1905 paper. The analytical results presented here in the various chapters and appendices indicate that certain of our technologies are far from optimized or understood, such as particle accelerators. The terminal speeds of accelerators have not been optimized nor are their maximum speeds being attained. Furthermore, the terminal speeds are not as high as are being used in the analyses of the measured data on the kinematics of particle collisions. If the analyses within this book hold, we have been grossly overestimated the electron and charged particle speeds that can be produced. Luckily, such assertions are easily tested using modern electronics and particle accelerators to fire charged beams down tubes of varying temperatures to assess the terminal speeds that are attained.

The key point is that the totality of the analysis supplied in this book supports the contention that relativistic kinematics are a misrepresentation of the real physics. We can duplicate, enhance, and advance the physics of relativity without any reliance on kinematics by simply using the electromagnetic theory first described in Einstein's 1905 paper. The key physics is identified by applying the physics of radiation pressure within various scenarios, in which new physics and limitations to associated technologies can be identified. While the key element is the Doppler expression, at higher velocities the aberration becomes important. We recognize the necessity of including the aberration in high-energy inverse-Compton scattering, though we missed the role of the aberration in properly quantifying Doppler measurements.

While it is true that for many practical purposes we can ignore the Doppler and aberration impacts on radiation pressure, we cannot ignore the impact in those cases in which what we measure are essentially perturbations on the first-order and classic physics. In fact, modern physics has driven us to better understand such perturbations in a wide variety of measurements. It is incumbent upon the physics community to understand the origin and magnitudes of these perturbations, because as technologies evolve, these perturbations do not stay simply perturbations. In order to avoid the trap that Einstein fell into with regard to relativity, we need to understand the basic physics well enough to distinguish between ignorance of something old before inventing something new that may simply be a symbol of our ignorance.

For a century, we only consider static radiation pressure models, and, consequently, we have totally missed the dynamic radiation model suggested by Einstein and how that might affect the acceleration of a mass or particle. On the other hand, the Doppler model derived in that 1905 paper had no more solid experimental foundation than did the totality of relativity. Consequently, either approach could have been adopted some twenty-five years later to explain the operation of particle accelerators. However, it is also clear that no one ever went back to that 1905 paper and really tried to understand what was being described.

As a final note, there are several ideas interwoven in the Electromagnetic Part of the 1905 paper. Even though the radiation pressure model exists in the 1905 paper, it was not used in the kinetic energy description to model how a particle would be accelerated via an electric potential. If not accelerated via a radiation pressure described as a wave phenomenon or as a photon flux, how else is the electron accelerated? Not enough was understood about photon theory at that time.

Einstein's Relativity: Some Conclusions

Using a mixture of electromagnetic and kinematic concepts, several ideas have entered the popular understanding of what Einstein was really doing in the 1905 paper. The popular result that $E = mc^2$ and the concept of a rest mass energy comes not from Einstein's relativity paper but from Minkowski's reformulated approach to Einstein's 1905 paper. Einstein's supposed mass-energy relationship in the 1905 paper had no rest mass and was simply a new expression for the kinetic energy. This formulation happened to include the expression mc^2 because of a conversion from velocity to the parameter β. If not for that, there would be no mc^2, a representation that has attained mythical status.

It might be argued that another of Einstein's 1905 papers, "Does the Inertia of a Body Depend on its energy Content?" also published in the <u>Annalen der Physik</u> and available in a translation in the same Dover book as the relativity paper, may suggest that there is a rest energy. However, that paper contains the same mixing of kinetic and electromagnetic factors and is no more reliable than the addition of velocities. In fact, Einstein consistently used Doppler expressions to describe kinematics, which are demonstrably wrong. Thus, the mass-energy relationship is another of those ad hoc expressions used to explain the unexplainable.

There was never an intention that we might be able to convert mass into energy nor does the 1905 paper suggest such a thing and Einstein never considered this. It was someone from the early 1930s working in the new nuclear physics who was trying to find a way of accounting for the apparent mass losses that occurred during certain nuclear decay events. Such identification is a natural outcome of photon theory. Since the momentum of a photon was measurable, we had an entity traveling at the speed of light that exhibited momentum, p. Hence $p = m^* c$, where m^* is a pseudo and non-inertial mass. Therefore, for a photon $E = h\nu = pc = m^* c^2$. The notion of a pseudo non-inertial mass was convolved with the notion of real inertial mass, which is the likeliest origin of the notion that mass and energy are equivalent.

However, there is a phenomenon called pair production in nuclear physics. It was first observed in ~ 1928 that a photon above a threshold energy scattered from a large nucleus could generate a positive and negative electron simultaneously. The threshold energy of the photon was related to $m_e c^2$, where m_e is the so-called rest mass of the electron. The physics does suggest that matter and energy are equivalent. However, as pointed out in Appendix 3, the form mc^2 arises out of the conservation laws when we scatter a photon off an electron and uses the momentum of the photon as h c/λ and energy of the photon as h ν. We do not actually have to have a rest energy of the electron, since as we showed in Chapter 3, there is no relativistic form for the kinetic energy of the recoiling electron. The form for m c² is a result of simply using classical physics. It is the same as when Einstein converted from using v to using β when he derived his kinetic energy expression, and the form m c² emerged from his mathematics.

And, as a follow up to the essay by Frank and Gleiser, the supposition within the physics community is that all the elements of the relativistic electrodynamics from the 1905 paper are known and complete because of the inclusion of a relativistic vector potential, a single expression, within the general theory. This is a case of mathematical elegance gone awry, since, at some point, one must actually posit a scenario and then solve the equations for the variables of interest. The equivalent perspective would be that having the general relativistic equations is equivalent to having all possible solutions to it. Furthermore, the notion is that we do not then have to find the solutions, because we have the master equation. In fact, as pointed out by Brillouin, it is very difficult to find solutions to Einstein's general relativistic equation for arbitrary scenarios, because the equations are so general. And, more egregiously, there turn out to be multiple possible solutions, since the equations are so general.

The general conclusion is that we can replace the "classical" interpretation of special relativity with an electromagnetic formulation. Such a replacement then allows us to refine and improve the various technologies based on the kinematical formulation. We lose nothing but gain much by the substitution. And in moving further into the domain of quantum mechanics, we may gain new insights into what, exactly, quantum mechanics may be through our new understanding of Newtonian and relativistic physics. In particular, we may finally have new insights into what we call field theories which rest on the foundation of classical relativity and old Newtonian "fields". But more importantly, if the analysis in this book and the one on Newton's gravity hold, we can legitimately challenge modern theoretical physics as suggested by Frank and Gleiser. The mathematics is not the physics nor should we simply dabble or doodle until "something interesting" emerges.

That said, I still wonder where the idea for the Electromagnetic Part of the 1905 originated if not as an element of somewhat idle speculation, though by whom? Lorentz? Poincaré? Einstein? The clue to the answer for this question lies, I think, in the paper by Logunov and the comments by Darigol plus the papers by Lorentz published in the Dover book, *The Principles of Relativity*. Many people came close and all contributed, but it remained for Einstein to make the crucial connection, even if it was accidental. Einstein got the credit but for the wrong reasons and for the wrong emphasis.

There is also a short discussion in Appendix 2 relating to the aberration. Einstein's aberration makes no sense without considering a measurement scenario. Einstein never included measurement scenarios directly, yet his mantra of stating that the observer sees this or that implies that he knew we needed to make measurements to know anything. The aberration predicts an interaction of photons with moving matter, yet that was never an element within his modeling. The astronomic angle of arrival does not require an interaction for the aberration to occur, since it is a geometrical phenomenon. Yet this perspective is exactly what Einstein' perspective was…the aberration occurs only because of motion. One can conclude that Einstein's approach to the Doppler and aberration accidentally predicted unknown physics. How did that happen?

But, without Einstein's 1905 paper and Einstein's relentless efforts at selling it, relativity would have become another obscure theory that might only have emerged years later as something else entirely, such as Einstein's electromagnetic theory that I have championed in this book. In essence, relativity was not ready for prime time as first developed. Furthermore, the only real and practical aspect of relativity was solely developed by Einstein.

Appendices

Appendix 1—Dynamic Radiation Pressure Models

This appendix is a more detailed extension of the discussion begun in Chapter 5 to include the impact of the aberration as discussed initially in Chapter 6. Also, this appendix had initially included an analysis and discussion of current research into solar and radiation sail technologies. I subsequently removed the systems and propulsion aspects of the discussion for inclusion as an appendix in the next book, *A Novel Propulsion System*. The reason was that the next book is a technology book whereas the current book is a physics book. Therefore, the material in this appendix is limited to analyzing the impact of the Doppler and aberration on radiation pressure.

To frame the discussion of the physics, I will use a limited number of scenarios, just as in Chapter 5, where the discussion expanded to include the impact of radiation pressure on the operation of particle accelerators. For the most part, we must invoke "science fiction" hardware and scenarios to even contemplate doing something useful with radiation sail technologies. In this appendix, I add to the existing radiation sail dialog by identifying a wide range of un-modeled and un-discussed physics associated with radiation pressure effects. Some of these new effects might impact existing and planned spacecraft missions. In addition, the physics introduced in the various scenarios has applicability to other non-spacecraft scenarios and systems, such as on particle accelerators, which was partially discussed in Chapter 5.

First, the details in this appendix are not meant for the faint of heart. Unlike a refereed professional paper, this appendix is long on description and detailed mathematical models. The reason is that there is a large host of unrecognized and unmodeled physics that will impact the more spectacular radiation pressure propulsion dynamics. We have already described some of these physics in Chapter 5 in describing how we have totally missed how particle accelerators work and how they are constrained. Consequently, this appendix is a showcase for a similar host of unrecognized physics.

For those with a background and interest in the current state of the art in radiation sail systems modeling, which includes a discussion of the physics of radiation pressure modeling as currently practiced, there are many references

Appendix 1—Dynamic Radiation Pressure Models

that supply an excellent introduction into the physics and technologies of traditional radiation pressure applications. Still other references expand these analyses into much more speculative modeling. What distinguishes these efforts is the anticipated range of the attained velocities within the scenarios. Most velocities fall into the small-β range consistent with radiation pressure effects associated with planetary missions. However, the more speculative scenarios are associated with interstellar propulsion and we are beyond the small-β approximations.

For traditional and small-β scenarios, *The Starflight Handbook* by Mallove and Matloff and the book *Solar Sails* by Vulpetti, Johnson, and Matloff are excellent introductions into the physics and technologies for radiation pressure propulsion. For large-β and speculative radiation pressure propulsion, the following are excellent introductions: Forward—"Starwisp: an Ultralight Interstellar Probe"; Landis—"Advanced Solar- and Laser-pushed Lightsail Concepts"; Lubin—"A Roadmap to Interstellar Flight"; Siegel—"Sorry, But Lasers Won't Get You to Mars Anytime Soon"; Popkin—"What it would take to reach the Stars?" Popkin's article is a popularized summary of the latest studies into realizing interstellar propulsion and missions, which is called Starshot, which builds on Lubin's radiation-sail concept. What these discussions all have in common is that they are describing scenarios in which the modeling employs the static radiation pressure model. In this appendix, we discuss how the Doppler and aberration supply perturbations on these analyses and what the magnitudes and impacts of these perturbations on the missions' modeling may be.

In most scenarios, the radial speeds of solar sails are so small that we can rely on the small-β approximations to radiation pressure to supply useful models. I use the term radial to represent the velocity component directly away from a source of radiation. Lateral is likewise used for motion perpendicular to the radial component of flux from a source of radiation. Often the angular or transverse motion is so small that the aberration effects are likewise small, though they are accumulative and so cannot really be ignored for long-duration missions.

As a solar sail or spacecraft gains more radial speed, it might be assumed that the aberration effects go to zero. However, in almost all scenarios, unless the radiation sail is purposely canted to supply some lateral force for slowing the orbital motion of the sail, some orbital motion is usually always part of the velocity vector for the sail. Even though direct radiation or solar radiation accelerations are affected by the distance of the sail from the radiation source, any lateral orbital speed is always present unless cancelled. However, in long-duration missions, the aberration supplies a force that does slow the lateral speed, which is discussed later. None the less, as the direct flux magnitude drops as the range from the source increases, the magnitude of any aberration retro-force likewise diminishes at the same rate as the direct radiation intensity drops.

If the aberration were to become an issue in solar or radiations sail performance, we might manage the launch of the sail to coincide with the greatest velocity vector toward some target, but since mission trajectories are almost always spirals, there will almost always be some lateral motion. None the less, while solar sails must spiral into and out of gravitational wells, radiation sails using artificial radiation sources as the driver radiation have sufficient acceleration to be able to power directly out from gravitational wells. In any event, any aberration effects from residual lateral motion can be managed if recognized as the source of apparent trajectory perturbations.

Without repeating the full development shown in Chapter 5, when we put the dynamic elements together, we have a generalized dynamic radiation pressure model: $P = P_o (r_o/r)^2 (1 - \beta \cos\varphi)^2 / (1 - \beta^2)^{1/2}$, where φ is the angle between the direction of motion for β and the direction of propagation of the flux. For radiation sails, $\varphi \sim 0$ for motion away from the radiation source, and the dynamic pressure model reduces to $P = P_o (r_o/r)^2 (1 - \beta)^{1.5} / (1 + \beta)^{0.5}$, since all perturbations on the Doppler related to the aberration from the direct acceleration are small, though any residual transverse or lateral motion may still be present. Depending on whether the incident and Doppler-shifted radiation is reflected or absorbed, we would also include a numerical factor of value between one and two to indicate the accelerating force resulting from the incident and Doppler-shifted radiation, where force per unit area of sail is P/c.

The inclusion of the aberration does not invalidate the first-order use of the direct radiation pressure in determining how much radiation pressure a flux of radiation supplies to a moving object. The aberration causes a lateral perturbation to the forces supplied by the direct pressure by adding momentum to the transverse component of the photon momentum at the expense of the lateral motion of the object or sail. In fact, the magnitude of the aberration depends on the direct Doppler shifted frequency of the incident radiation. This Doppler shift is used mainly to quantify the momentum transfer and radiation pressure producing the radial motion of the sail.

To quantify the above assertions, we need to consider the impact of the aberration on sail dynamics. The analyses also hold for any spacecraft subjected to solar radiation, and, consequently, we use the Doppler and aberration models

Appendix 1—Dynamic Radiation Pressure Models

for radiation incident on a single surface. To first order, the incident angle of radiation onto a surface does not change the aberration effects. As discussed in Chapter 6, the aberration is caused by the lateral motion of electrons, either within a solid or as a spatial current, as the electrons move through a flux of photons, where the momentum of the electrons is exchanged with the photons. In direct radiation pressure, the photons give up their momentum to the electrons in a solid causing a pressure, and reflection doubles the effective momentum transfer. The same occurs with the aberration, and the incident flux is Doppler shifted depending on the radial velocity of the sail relative to the incident flux, and reflection would double both the direct pressure and the lateral momentum transferred to the reflected photons.

Going back to some radar scenarios used previously, if a stationary radar paints an aircraft moving with lateral relative motion, the reflected radiation is Doppler shifted by the radial component of the relative motion, but we also have an aberration effect that slightly upshifts the reflected radiation and aberrates the reflection or scattering angle. We can see for aircraft that these two effects are miniscule and undetectable, certainly for radar systems. Consequently, both effects exist but are essentially irrelevant to the practical aspects of the radar scenarios.

We also discussed that the detected radiation would also be Doppler shifted and aberrated by the hidden motion of the "stationary" radar and, as previously stated, there is no practical impact from any of these motions. Thus, radiation reflected from radiation sails both has a small blueshift on the reflected and Doppler-shifted radiation and is aberrated away from the specular reflection direction. It is only long-mission times that allow the small aberration effects to accumulate in the motion of a radiation sail. (Even though the aberration increases the transverse momentum of reflected radiation, this minor increase in energy is usually swamped by the simultaneous Doppler shift, which we quantify later.)

Using the small-β approximation, we can quantify the impact of the aberration on the motion of an object. If we have an incident flux on a moving object, the direct Doppler effect shifts the incident power that hits the object. This same Doppler-shifted radiation then is aberrated away from the specular reflection direction. Whether the radiation is absorbed or reflected only changes the momentum transfer by a factor of between one and two, which also holds for the aberration regardless of the incident angle of the radiation on the object.

If the velocity of the object relative to the flux direction is at some angle φ, then $v \cos\varphi$ is the direct Doppler shifting velocity and $v \sin\varphi$ is the lateral velocity. Therefore, the deviation of the radiation after interacting with the object is $\delta \sim \beta \sin\varphi$, which we derived earlier in the book, where, for now, we ignore the residual lateral velocity of the sail. For the direct radiation, the angle φ supplies a new incremental lateral momentum added to the Doppler shifted radiation. If the incident flux is P_o, then for total reflection, the reflected flux is $2 P_d$, where P_d is the Doppler shifted flux. The flux P_d is aberrated by angle δ, which means that the new reflected flux has an additional momentum component given by $2 P_d \beta \sin\varphi$. Since the direct radiation penetrates the sail and is reflected from the electrons, there are twice the opportunities for lateral momenta changes. It is likely that the aberration requires the magnitude of $2 P_d$ to find the correct magnitude of the aberrated and reflected flux, so in what follows, we use a flux of $2 P_d$ in determining the magnitude of the aberration on the motion of a surface moving laterally through a flux of radiation.

Consequently, the effective force accelerating the object comes from both the direct Doppler shifted flux, calculated using the dynamic radiation pressure model, plus a transverse retro-force given by $2 P_d \beta \sin\varphi$. We can see that if the motion is mostly radial, $\varphi \sim 0$, and the lateral retro-force is also ~ 0. For a solar or radiation sail with mostly a transverse velocity through the radiation flux, $\varphi \sim 90$ deg, and the new vector force on the object is the static radial force that is perturbed by an orthogonal component that is $\sim \beta$ times the direct radial component of pressure and is in the opposite direction to the transverse motion. The key point is that both the Doppler and aberration supply retro-forces on the static radiation pressure accelerations.

In quantifying the aberration, we need to also be sure that we are using the value of β associated with the transverse velocity. Normally we do not make these velocity component distinctions and simply refer to the speed of a space object. Since we have not incorporated Doppler and aberration effects into the radiation pressure models up to now, there has been no need to make these distinctions. We now have a radial velocity component that determines the radial pressure which also determines the magnitude of the radiation flux that is then subjected to the aberration based on the transverse velocity component. The radial driving pressure is independent of the lateral pressure but not vice versa.

In the small-β approximation, the direct radial force per unit area of an object arising from the dynamic radiation pressure for a solar sail becomes

Appendix 1—Dynamic Radiation Pressure Models

$F_d = 2 P_o (r_o/r)^2 (1 - 2\beta_r)$. Therefore, the static radiation pressure is decreased by a small amount, which is a perturbation of magnitude $\Delta F_d \sim - 4 P_o (r_o/r)^2 \beta_r$, where β_r is the radial velocity component. The aberration force has magnitude found from $P_a = 2 P_d \beta_t$, so that we have $F_a = 2 P_o (r_o/r)^2 (1 - 2 \beta_r) \beta_t$, and the actual perturbation, to first order in β, becomes $\Delta F_a \sim - 2 P_o (r_o/r)^2 \beta_t$. Therefore, consequences of the Doppler (plus compression) and aberration as perturbations on the static radiation pressure model are two retro-forces that are, to first order, slowing a radiation sail, or any object, in both the radial and lateral directions relative to the static radiation pressure force on the object. These two retarding forces are acting rearward at some angle ψ relative to the negative radial direction, which is opposite the flux direction. The angle is given by arctangent $(\beta_t/2 \beta_r)$ and the magnitude of the total retro-force is $\Delta F_{total} \sim - P_o (r_o/r)^2 \sqrt{4\beta_r^2 + \beta_t^2}$ per unit area of an object. In developing these formulae, we have only kept terms to first order in β, which holds for motion within the solar system. Since all conceivable values of β in the solar system are < 0.0003, we see that the total perturbation is an additive negative force on the static radiation force.

It should be noted that the aberration used by astronomers was not developed for the aberration of flat surfaces, only for imaging devices and the aberration of an image. Only Einstein's aberration model is suitable for discussion of the aberration from single surfaces. In Chapter 6 the mnemonic of using the shift of an image point is clearly a fabrication that, while useful to visualize what is happening in an imaging device, this mnemonic is not truly useful in discussing the physics of the aberration from single surfaces.

In general, we can say that all static radiation pressure results will be perturbed by the amount described above for any object. However, the aberration can be reduced or eliminated in many scenarios and either managed or ignored to a certain extent, which we will do in the remained of the radiation-sail discussions in this appendix. None the less, in a more practical world of radiation pressure effects on spacecraft and satellites, we would need to retain both the Doppler and aberration, since, while small, these effects accumulate over long-duration missions, especially for those trajectories into the inner solar system, whether for gravitational assist or solar orbit missions. Also, in planetary scenarios, the lateral velocity remains significant over the total mission.

Taking the discussion off track for a few paragraphs, consider a spinning and translating symmetrical object embedded within a radiation flux. This might be a spin-stabilized cylindrical satellite in orbit. The instantaneous velocity of any point of the spinning surface subjected to an incident flux is a vector sum of the translation and spinning velocity vectors relative to the flux direction, and the forces from the aberration became asymmetrical across the object. Therefore, the object's spin is slowed, and the total object also experienced a small lateral force from the aberration slowing the object's transvers or lateral (orbital) motion. The lateral force is to first order essentially the same magnitude as would occur on a similar stationary surface moving laterally through the radiation flux. The components of force on the object are additive. Consequently, the direct radiation pressure acceleration is modified by the small lateral acceleration. While these various forces are deterministic, they can result in complex dynamics. We ignore these complexities and view the sail as a static flat surface in everything that follows.

So, let's summarize what we will be working with. The forces caused by the direct and aberrated radiation can be easily quantified to first order for sails because these are essentially static surfaces that maintain a fixed area and orientation in relationship to the incident flux over protracted periods. The direct flux is Doppler shifted by the radial motion, and a compression factor is calculated given by the instantaneous radial velocity. This same compressed and Doppler shifted radiation flux is used to calculate the lateral forced caused by the aberration. The sail is reflective, though whether absorptive or reflective, the only impact is a numerical factor that is greater than unity but less than or equal to two, where two is for perfect reflection. The direct flux drives both the radial and lateral retarding accelerations, though the magnitude of the lateral acceleration also depends on the lateral component of motion.

As a solar sail leaves Earth's orbit and is pushed outwards, the path of the sail becomes a spiral, first about the Earth and then about the sun. Some small Doppler and aberration effects would occur as the sail orbits the Earth but would change signs and be "back and forth" as the velocity of the sail or object relative to the sun oscillates with each revolution about the Earth. Only over long integration times would there be a measurable difference between the static and the first-order dynamic radiation pressure models.

As the sail leaves Earth's orbit, its radial velocity starts near zero and increases as the sail moves outward and into a higher orbit. An apparent paradox of orbital mechanics is that as the radial distance increases, the total speed is reduced until an object achieves sufficient distance from its primary, such as Earth, that the radiation acceleration is now sufficient to begin to gain

Appendix 1—Dynamic Radiation Pressure Models

radial speed and increase the distance as the object achieves its escape velocity. The usual way of changing orbits is to increase the orbital speed such that centrifugal forces increase the outward radial speed. Therefore, an effective use of a sail to gain radial speed is to cant the sail to achieve greater lateral speed, such that the sail spirals out of its orbit and becomes free of the primary.

A space mission inward towards the sun, in which radiation pressures increase rather than decrease, would show the most deviations between the static and dynamic solar pressure effects on the spacecraft's trajectory, because of the increased flux and speed as the range to the sun decreases. For missions outward away from the sun, where the radiation pressure decreases, the impact on speed would be less pronounce despite the mechanisms, such as gravity assists, used to accelerate the spacecraft to speeds as much as twice the assisting planets' orbital speeds.

The scenarios can become complex as we look at relative angles, ranges, and velocities. For instance, a gravitational assist from the inner planets and again from Earth could supply a radially outward velocity more than twice the Earth's orbital speed plus the existing speed of the object. As the sail leaves Earth's orbit, the angle φ could be close to zero and $\beta \sim 0.0002$. Using this value, which we assume was acquired from a gravitational assist, we find that the first-order dynamic correction, which includes the relative speed compression factor plus the Doppler shift, only accounts for the correction factor $\sim (1 - 2\beta) = 0.9996$ for twice the Earth's orbital speed in the radial direction. In other words, the retro-force is a perturbation of $\sim -2\beta$ times the static radiation pressure driving force. If the sail simply accelerates away from the Earth without an assist, the Doppler correction factor would be essentially zero until sufficient radial speed is attained, if possible. In this scenario, however, we have a lateral value for β, and $\beta_t \sim 0.0001$. In this scenario, we might also have a retro-orbital force of magnitude given by $\sim -2\beta_t$ times the static radiation pressure, since there is no Doppler shift and, therefore, no Doppler and compression perturbations, and the static model holds for radial acceleration.

When we dynamically integrate the force or motion equation for the sail slowly spiraling away from the Earth, we also include both the gravitational attraction of the sun, which is slowing the sail, and the inverse range decrease in solar flux as the sail moves radially farther from the sun. The net effect is for the dynamic corrections to have no appreciable effects for currently attainable solar sail scenarios, other than to be the source of dynamic perturbations for which no explanations might otherwise exist. And, in fact, such small perturbations are regularly experienced by long-term space missions.

We are omitting any discussion of a hidden velocity. For just the sun orbiting within the galaxy, the radiation from the sun would show a Doppler shift depending on the direction of the flux relative to the orbital motion. If this Doppler-shifted radiation then encounters a moving object, that object's motion is not just relative to the sun but includes the hidden motion. In this simple linear model, the sun's motion is common and there is no net Doppler shift. Consequently, the effective radiation pressure is all due to the radial motion of the object relative to the sun. As some of the scenarios are extended, we must be more careful in making such assumptions when, for instance, an object leaves the solar system and approaches another stellar system in which the galactic component of the hidden velocity is different from that of the sun.

The escape velocity radially outward from Earth is ~ 7km/sec, but the comparable escape velocity from the sun from Earth's orbit is ~ 42km/sec. Small sail thrusts alone likely will not allow an object to spiral away from Earth into a solar orbit without additional lateral speed. This same issue exists for a solar sail attempting to spiral away from the sun. The value of a gravitational assist is to acquire both transverse and radial speeds, otherwise a "sailing" object cannot spiral out of the solar system with just the solar radiation pressure supplying the outward thrust, using standard sail technologies.

Only the laser sources described by Lubin, which are discussed next, allow the inexorable laws of orbital mechanics to be bypassed. None the less, because the speeds are all small-β for current solar-sail concepts, the static radiation pressure adequately models the gross dynamics of an object undergoing solar-pressure accelerations. It has been known for as long as people have been performing solar-sail calculations that solar energy is simply insufficient for practical propulsion, because the sails are too heavy.

Lubin describes a laser-driven system that was adopted, in part, for the new interstellar mission called Starshot. Lubin's initial efforts are directed toward feasibility studies for launching fleets of small sensor packages for missions throughout the solar system. The Starshot mission has many features in common with the solar-system missions. However, the interstellar mission is essentially an update to an older study by Robert Forward. Forward described an interstellar mission using a microwave-driven system called the Starwisp, which was first described in 1985 and elaborated on by Landis in 2000. Forward

Appendix 1—Dynamic Radiation Pressure Models

hypothesized a microwave beam system that employed a variable focus microwave Fresnel-lens transmitter to maintain the beam power totally on the sail for as long as possible until the physical optics no longer allowed the beam spot to be as small as the radiation sail. This strategy allowed the radiation to totally fill the sail out to some critical distance without any radiation spill over. Landis revisited the Starwisp concept in 2000 and discussed various technological issues as well as the evolution of technologies from microwave frequencies up to millimeter wave frequencies. It was at this point that Landis also commented on and then dismissed the existence of the Doppler as a possible influence in such a system's performance. And, as an inevitability, the driver radiation has been further modernized to include a solid-state laser phased array for a modernized radiation sail mission.

Still, as Popkin describes the efforts, we are looking at a 1km squared cluster of phased-array solid-state lasers, which supplies the same drive-powers and mission parameters as modeled by Forward using a 50,000km diameter microwave maser transmitter supplying driver radiation. These scenarios are all highly speculative, but the idea is to send a sensor package to a newly-discovered planet in the Proxima Centauri system in a single lifetime, such that reconnaissance information can be beamed back to Earth. The scaling to a laser source was accompanied by a comparable increase in total source power and a concomitant radical decrease in the radiation sail size and mass to ~ 16 m^2 in area with a total mass ~ 1 gm, which can be compared to Forward's 1km square sail with a total mass of 20 gm, including the payload. Such scenario scaling can lead to almost any performance. These scenarios have something in common in that they envision an attained speed of ~ 20% the speed of light within the solar system, though the system described by Popkin is scaled to achieve this speed within two up to ten minutes of irradiation, depending on which version of a summary about the Starshot mission is being read.

What makes these scenarios of interest here is that the attained speeds are sufficiently high that the Doppler perturbations are no longer perturbations but crucial features in modeling the dynamics of these systems. While not particle accelerator scenarios, these scenarios allow many of the features of the particle accelerator physics to be explored at lower speeds than occur in particle accelerators. However, the inclusion of ambient radiation effects may be a forced effort, because from a practical perspective, Lubin points out that we can design these radiation sails to essentially be immune to the effects of the ambient radiation, at least to a limited extent.

On the other hand, Lubin's strategy assumes that the radiation sail can be tailored to be immune to any ambient radiation other than that of the laser-driver frequency. Dielectric or dichroic mirrors are wavelength sensitive, and the Doppler effect changes the wavelength in these scenarios by a sufficient amount to raise questions as to how effective these mirrors would be across the total expected Doppler-shift region. For instance, the Doppler shift of an ytterbium laser is shifted from ~ 1000nm to ~ 1200nm during the acceleration period of the Starshot scenario. This requires that the dichroic mirror actually be a quasi-broadband mirror. Some solar radiation would be downshifted into this same band, so that the sail would not be immune from some solar radiation effects, small as they may be, which are modeled later in this appendix.

Once Starshot reaches interstellar space, the peak ambient radiation from the Alpha Centauri system becomes upshifted into the blue or even ultraviolet region of the spectrum. Therefore, the characteristics of a dichroic mirror…or any selective mirror…would need to supply a very broad transparency while still allowing efficient laser reflectance in the presence of the Doppler shift while simultaneously supplying immunity (transparency) from ambient radiation effects. That being the case, it is likely that the Starshot sail is, to one degree or another, a broadband sail with some spectral reflectance and spectral absorption characteristics that are yet to be explored. The best we can do at this point is to calculate the potential impact of the ambient radiation on the mission to account for these perturbations or to eliminate them from consideration.

The reasons Forward used microwaves are that a very low-mass grid of conducting wires suffices to create an antenna tuned for capturing the microwave radiation while allowing other radiation to pass unhindered through the mesh. Such a design is necessary to achieve the accelerations needed to reach practical speeds for interstellar missions. In the case of Starwisp, the mission was to send a sensor package to the Alpha Centauri stellar system located at 4.37 light years from the sun, since at that time the planet orbiting Proxima Centauri (Proxima b) was unknown. However, Landis concluded that the Starwisp as described by Forward was totally unrealistic and would be fried to a crisp, per Landis, at the first instant of applied power. Consequently, Landis looked at more realistic millimeter wave systems. Still, these are little more than science fiction descriptions at this point despite the high-quality analyses.

Once a radiation sail becomes broadband, meaning that it is influenced by optical radiation of many wavelengths, the mass of the sail increases enormously,

Appendix 1—Dynamic Radiation Pressure Models

because such sails require reflecting metallic or dielectric surface layers. The added mass of the layers on some light substrate lowers the attainable accelerations to impractical levels. Lubin's sails are likewise dielectric mirrors that are intended to be reflective only at the wavelength of a driver laser's wavelength and are otherwise transparent, a design that may be unattainable when the Doppler shift is included as a design boundary condition. The issue is that this is still a layered sail that is too heavy to be practical, except for the enormous irradiance levels envisioned for Starshot. Consequently, new fabrication strategies and materials will be needed to achieve practical radiation sail propulsion. Plus, the reflectance of the mirrors must be nearly 100% to avoid being, as Landis stated it, fried to a crisp at first power.

However, since Landis's analysis, graphene materials have emerged. Graphenes are single carbon layers with considerable lateral strength. Using graphenes to lower the substrate mass or even to reduce the active-sail area may help reduce radiation-sail masses to acceptable levels. If a graphene sheet could be thinned to be a mesh rather than a solid sheet, or if we could deposit a metallic grid on a graphene substrate, we could possibly achieve Lubin's goal of a wavelength-selective sail with almost no significant mass. Conductive meshes used as optical filters were first explored some forty years ago, when laser systems began to emerge as tactical sources of illumination, such as laser designators and, later, as counter-sensor weapons. Therefore, the graphene sheet would have much less areal mass density than anything currently under investigation.

Despite the unrealistic aspects for any of the scenarios, we will use the outline of Forward's concept to show how Starwisp might be affected by the ignored Doppler effects. The results are applicable to Starshot as well, in so far as the assumptions and analysis are parallel between these two scenarios. All the described radiation sail scenarios will be impacted by the Doppler effects in much the same way, since the scenarios require acceleration to ~ 20% of the speed of light. We ignore any aberration effects and note that any electromagnetic radiation would experience the Doppler effects described next.

The Starwisp was configured to be accelerated to ~ 20% the speed of light within the first week of irradiation and while still within the solar system. For the Starwisp and Starshot, the terminal velocity is $\beta \sim 0.2$. However, by introducing the dynamic correction factor, we find that there is a decrease in the effective radiation pressure to only ~65% of the static value at the termination of the first phase of acceleration for Starwisp. Thus, in the posited scenario, the terminal velocity would be smaller than could be achieved by Starwisp using the technologies posited by Forward, Landis, and Lubin for an interstellar mission. The Doppler's impact on the Starwisp is marginal during the initial hard acceleration phase of the scenario, but the cumulative effect is to considerably reduce the terminal velocity of the system. These factors are described quantitatively next.

A by-product for including the dynamic radiation pressure factors is that these factors modify the equations of motion considerably, such that the simple approach used by Forward for finding the terminal velocity does not work as it did for him. The acceleration, a, as calculated by Forward used a static radiation pressure driving a mass M, where M is a total mass of the system…sail plus payload. The acceleration at some distance r from the source is then given as $a = 2\, e_F\, P\, A\, (r_o^2/r^2)/c\, M$, where P is the flux or power per unit area onto a radiation sail, A is a square sail of area 1km squared in Forward's scenario, and the flux is given in watts per unit area at some reference distance r_o from the source, e_F is an efficiency factor of 0.34 because the sail is a grid of wires, c is the speed of light, and r is the attained distance of the object away from the radiation source. (In the presence of Doppler effects, e_F is also a function of β.) We are also assuming the angle φ is zero and that the object is a perfectly reflecting object, which is where the factor of 2 comes from.

If we add in the dynamic correction factors, we have a new dynamic radiation pressure acceleration model given as $a \sim 2\, e_F\, P\, A\, (r_o^2/r^2)[(1-\beta)^2/(1-\beta^2)^{0.5}]/c\, M$. Therefore, we only need a single equivalent $e_{F\,optical}$ in order to account for any possible optical interactions with the sail, including subsuming the reflectance factor 2. In the absence of the dynamic corrections, the first integral of the acceleration gives the velocity, and when the integral is taken over the range from r_o to infinity, we find a first order terminal velocity. We ignore gravity as did Forward, though gravity can be added as an inverse square term. Once we modify the technology to be that of a laser-driven sail, we can scale all the parameters together so that we have a comparable total flux and total system mass that matches that of Starwisp, much as employed with the Starshot modeling, and carry out the modeling efforts oblivious to what the driver radiation and sail dimensions might be. (It is a characteristic of these types of parametric analyses that we can always posit conditions that make the result of the analyses to be what we want them to be. In the case of Starwisp and Starshot, we posit impossibly powerful sources of driver radiation fluxes to attain the required mission profile.)

Appendix 1—Dynamic Radiation Pressure Models

Forward's scenario is divided into two dynamic portions. The first portion is the initial phase in which a dynamically focused microwave radiation source can supply a constant acceleration that has no inverse distance dependence on the radiation intensity. This phase lasts until the sail reaches a reference range r_o, which is defined by the range at which the focused microwave beam begins to overfill the solar sail. When we add the dynamic terms into the radiation pressure model, the acceleration is now velocity dependent and not constant. The above form is fortuitously separable in terms of velocity and time, since $r = v\,t$ and $a = dv/dt$.

Alternatively, by noting that $dv/dt = (dv/dr)(dr/dt) = v\,dv/dr$, we can arrive at a somewhat simpler form than if we look at time. Making these substitutions in the acceleration equation given above, we find we have two simple-looking algebraic equations to integrate, one for the constant acceleration phase of the scenario and a second equation for when the intensity of the radiation begins to fall off as an inverse-squared range term:

$$\{\beta\,[(1 - \beta^2)^{0.5}/(1 - \beta)^2]\}\,d\beta = K\,dr/c^3 \quad \text{(constant flux)}$$

and

$$\{\beta\,[(1 - \beta^2)^{0.5}/(1 - \beta)^2]\}\,d\beta = K\,dr/c^3 r^2 \quad \text{(inverse-square-range flux)},$$

where we use the fact that $v = c\,\beta$ and $dv = c\,d\beta$ and where K is the static acceleration from the Starwisp non-dynamic pressure models for the scenario and is the acceleration used or found by Forward inputting his assumed sail size, the total system mass, the sail efficiency or cross section factor, and the driver source output flux. Unfortunately, the solutions to the above equations are transcendental, and the results for specific scenarios must be graphically or numerically depicted as velocity versus range. In addition, Forward's sail efficiency factor e_F will also be affected by the Doppler, which should actually allow the efficiency to improve as the frequency decreases and the wavelength increases, the effects are not as pronounced as for an optically reflective sail.

In the following modeling, we use the above two equations plus another one for finding accelerations with various input fluxes that represent various pieces of the expanded scenarios described in this appendix, which incorporate a variety of ambient sources of flux that will perturb the primary scenario that only considers an artificial driver radiation source. The additional and external fluxes are from the sun and from the Alpha Centauri stellar system plus other ambient stellar fluxes acting on the modified Starwisp. These fluxes variously push against the modified Starwisp either to accelerate it or to retard its motion. In addition, gravity is also considered in retarding the acceleration of Starwisp, though we "add" each source of perturbation one at a time to assess the impact on Starwisp one at a time. This approach is possible since these are small effects that are only additive perturbations.

The reasons for developing the hybrid system are three-fold. One reason is simply to showcase and quantify the unknown physics. The second reason is that in the next book on a novel propulsion system, we develop a model for an interstellar vehicle which is a solid object subjected to all the ambient constraints that will exist. The third reason is that, while Lubin discusses some mitigating strategies to make his sail immune to any radiation other than his driver radiation, and a grid-like radiation sail can mitigate the ambient effects to some degree, it is useful to understand how the constraints impact the hybrid system, so that we recognize the existence of these constraints and include them or exclude them as the modeling indicates.

In the models, we look at the same mission as Landis and Forward in finding accelerations, velocities, and flight times along various segments of the mission profile. In making these calculations, we use the same three models described previously with some variation and with various flux levels to simulate the direct and indicted effects of the various ambient fluxes on the mission profile. The purpose is to demonstrate the impact of the dynamic factors in modifying the static radiation pressure and how these factors modify the mission profile of a Starwisp-like hybrid system.

Forward's system designs and scenario have a feature that we can use to more easily quantify the impact of the various dynamic factors. That feature is that the initial constant acceleration phase lasts so long that the Starwisp reaches a velocity $v_o \sim 13.2\%$ the speed of light, or $\beta_o = 0.132$, at the transition range r_o, at which point the dynamic model transitions to an inverse-square acceleration model due to driver-beam divergence. The dynamics in Forward's model show that at the transition range, $v_0 = 3.96 \times 10^7$ m/sec after 35,000 seconds (~10hrs) of acceleration out to range $r_o \sim 6.8 \times 10^{11}$ m, which is roughly the distance from Earth's orbit to just past Jupiter's orbit. The steady state, static, or linear acceleration K is 1130m/sec^2 over this first segment of the trajectory.

Forward's magnitude of the acceleration combines the system values calculated from the combination of power source dimensions and output

Appendix 1—Dynamic Radiation Pressure Models

power, the mass of the sail plus payload, and an efficiency factor for how efficient the sail was in scattering the incident radiation. This is because a grid has a totally scattering cross section dependent on the radiation wavelength, on wire conductivity, on grid wire sizes, and on grid dimensions. For the first phase of acceleration, we integrate each side of the dynamic radiation model separately and use the boundary condition that at $r = 0$, $v = 0$. From this, in principle, we solve the initial phase equation for the new dynamic model to find the new values of v_o and t_0, where the transition range r_o stays the same. (Since we use r_o as the flux transition range, $r = 0$ is at Earth's orbit. Later when including the solar flux, $r = r_E$ which is a reference range and r becomes the distance from the sun and not the Earth's orbit.)

However, the solution is transcendental. Using Mathematica®, we can easily numerically integrate the dynamic model for the steady-state acceleration. Using an iterative process, we find that the terminal distance is now reached in 36,483 seconds rather than in 35,000 seconds. We further find that the velocity at transition is now 12.1% the velocity of light rather than 13.2%. Consequently, while the initial phase of the scenario is only somewhat impacted by the dynamic radiation pressure terms, there is an impact.

However, using numerical integration over the rest of the scenario using the second dynamic equation developed previously, the integral is now from the transition range out to infinity. The result is that the terminal velocity is now ~16.6% the speed of light, which is considerably less than found from the static model and which Forward had calculated a maximum velocity of 20% of the speed of light. Such behavior is similar to other types of models in which speeds of less than ~10% the speed of light is involved, in which deviations from the classical models are small and are treated as perturbations from some unknown sources. So far, these results also hold for Forward's model.

At this point, we have found the major impact of the Doppler on the Starwisp mission and, likewise, on the Starshot mission. For the attained speeds in these scenarios, the Doppler effect was much larger than merely a perturbation as it would be for a planetary scenario. However, we now switch scenarios to allow ambient radiation to supply a perturbation on the above results. By the nature of these scenarios, the impact of the ambient radiation is a separate scenario that uses the same models as for the Starwisp and Starshot scenarios but is now more appropriate for the scenario of a starship with a solid-mass cross section.

However, the mass of a starship means that the only speed at which the ambient radiation might have an impact is when the speed is very close to the speed of light. This is because only at those speeds do the Doppler shift and compression factor cause the radiation pressure from the ambient sources to become comparable to the driving forces on such a massive object. At lower speeds, such as for Starwisp and Starshot, the small mass of these two systems allows the small ambient radiation pressure effects to have a measurable impact on the mission profile.

We can also find a terminal velocity by including external radiance incident on the forward surface of an accelerating mass. Normally, the terminal velocity is defined when the net accelerating forces go to zero. So it would be with a hybrid Starwisp. The analogy is with an object falling from a height toward the Earth, the viscosity of the air…or any fluid…supplies a resistance and a retro-force slowing the rate of fall. External ambient radiation can supply such a retarding force, though this same redshifted radiation flux behind the sail can supply additional acceleration. As pointed out by Lubin, if a driver is a laser source, the sail can be tailored to be reflective at that wavelength and transparent at all others. However, this may be only an inefficient implementation when Doppler effects are included in the analysis.

As a radiation-driven object or radiation sail gains speed, any small external radiation flux incident on the front of a sail experience both a Doppler up shift and a relative speed compression term, both of which increase the flux incident on the sail. Similarly, the flux from the rear direction experiences a Doppler red shift and a negative compress, both of which decrease the flux reaching the object. These dynamic effects have not been identified up to now in these scenarios. For instance, in the case of Forward's scenario, we find that the terminal velocity is ~16.6 % of the speed of light factoring in the additional "headwind" perturbations caused by the external ambient radiation. At 16.6% of the speed of light, the forward incident ambient radiation pressure from extra-solar sources is increase by an amount proportional to $(1 + \beta)^2/(1 - \beta^2)^{0.5} \sim 1.4$ but the rear pressure is decreased by $(1 - \beta)^2/(1 - \beta^2)^{0.5} \sim 0.70$. It only remains to calculate what the impact of these external ambient sources may be on the sail dynamics.

In general, considering a broadband radiation sail, light radiation from all the stellar objects within the hemisphere in the direction of motion of the sail is much larger than the value in the driver microwave freqiemcy to which the Starwisp was tuned. Since the Starwisp sail limits the radiation with which it interacts, the ambient flux would have a negligible impact compared to the impact on a broadband sail. However, we still need to quantify the effect to properly dismiss it,

Appendix 1—Dynamic Radiation Pressure Models

because this interaction does affect any object traveling at these speeds. In fact, a millimeter scale grid clearly has optical properties as anyone who has looked through a screen can attest. While optical radiation wavelengths are much smaller than the screen grid, the size of the wires and the grid size clearly have an impact on optical radiation, so the notion that the Starwisp sail is immune to optical radiation is a supposition yet to be quantified.

In the scenario of a mission to the Alpha and Beta Centauri system, which is a visually unresolved double star system, we have two stars with nearly the same color temperature as the sun but with a combined luminosity almost twice that of the sun. These two stars orbit a common barycenter with a high eccentricity, with Beta's distance from Alpha ranging from approximately the orbit of Saturn to that of Pluto. For this analysis, we ignore Proxima Centauri, which is a third member of the Alpha Centauri system and, as a small red dwarf, supplies insignificant radiation within the posited scenario. Therefore, the major Centauri stars plus the sun are the most obvious sources of ambient radiation in the scenario. Consequently, for computational purposes, the flux from the sun was doubled to represent the incident flux from Alpha Centauri and to represent to a reasonable approximation to the direct stellar flux incident on either side of the Starwisp sail.

Though there is other omnidirectional ambient stellar radiation, its contribution is much smaller with regards to radiation pressure effects compared to the radiation between the sun and the Centauri system incident directly on the rear and front of the sail, respectively. We will also simply assume that the radiation sail is 100% reflective on both sides of the sail for all optical radiation, though as Lubin discusses, we can tailor the reflectivity of each side of the sail to a certain degree.

Considerable data were gathered in the 1950s and 60s on cosmic electromagnetic radiation spectra, but for the Starwisp or a broadband-sail vehicle, we need the total integrated flux from the various cosmic sources and this data does not seem to exist. We can make an estimate based on the flux from our sun at the location of Alpha Centauri at 4.37 light years, or $\sim 4.3 \times 10^{13}$ km. Using the flux at Earth of 1353 w/m^2 at 150Mkm (one astronomical unit or AU) from the sun, the square of the ratio of distances yields a flux $\sim 1.64 \times 10^{-8}$ w/m^2 for all electromagnetic radiation flux from the sun incident onto the rear of the sail at the distance of Alpha and Beta Centauri.

Alpha and Beta Centauri together supply nearly the same spectrum but twice the blackbody radiation as our sun, so that they produce a flux on Earth of ~ 3.3 x 10^{-8} w/m². Recent work by Jensen, et al as reported in the 2001 SIGGRAPG Proceeding on modeling the ambient night spectral radiation on the Earth shows that the contribution by just the star light is ~ 3×10^{-8} w/m², the same magnitude as the direct irradiance of Alpha and Beta Centauri on the Earth. The analysis was made using the star catalog of known visible stars seen from the Earth and using the spectral temperature of each star, its distance, and a projected surface area to calculate the integrated irradiance.

Note that at Alpha Centauri, the microwave driver flux on Starwisp would still be supplying ~3 x 10^{-6} w/m², for a total of ~ 2 - 4w of total intercepted narrow band energy on a 1km-square antenna. The driver flux is still overwhelmingly larger than stellar microwave emissions, and the received power for Starwisp was deemed necessary to operate the posited on-board embedded electronics. Additionally, the Starwisp would continue to experience thrust even at interstellar ranges from the microwave transmitter and there would be no slowing flux from Alpha Centauri, so Starwisp was simply passing through Alpha Centauri without interacting in any way.

On the other hand, except for power to the electronics, the additional thrust is essentially irrelevant to the mission once the maximum speed of ~16.6% of the speed of light is achieved. There is no reason to keep the driver source on other than to supply ambient power during the mission, which may be a useful strategy. Otherwise, the driver power can be turned off after the nominally maximum speed is attained and the Starwisp coasts until the power is turned back on to supply on-board system power during the terminal phase of the mission. In a way, this is a ludicrous strategy, given that turning the maser source off to save power and wear-and-tear is a fiction on the same order as even positing a 50,000km diameter transmitter mirror or lens in the first place.

However, we can use the respective direct stellar irradiances from the sun and the Alpha Centauri systems to represent the only important ambient flux which would impinge upon Starwisp. This is a reasonable assumption, since either one or the other direct sources dominates the total incident flux along the mission trajectory. In addition, the indirect ambient flux is always a small perturbation on the direct fluxes at all points along he trajectory. Therefore, the forward and rearward pressures would be defined at any given speed β by the difference between the forward and rearward sail reflectivity and the difference between the redshifted and blueshifted and forward and rearward compressed radiation flux intensities. From this we

Appendix 1—Dynamic Radiation Pressure Models

calculate a net "headwind" or retro-force on the sail, if any. And, it is worthwhile noting that the direct flux for either the sun at Alpha Centauri or Alpha Centauri at the sun is essentially the same as the background stellar flux.

Also, to reiterate, the analysis is contrived to demonstrate the physics of several unknown effects which have been considered as negligible contributors to the dynamics of radiation sails. In some instances, this is true but in others it is not. The analysis does not consider gravitational effects, which have essentially doomed solar sails to fail without a gravitational assist. In addition, since the Starwisp velocities quickly approach ~ 0.1c, the small escape-velocity loss of velocity shows up in the fourth decimal place of the calculated velocities…which can be ignore. However, this small gravity loss in the velocity is integrated over the total trip, even though in the end it is essentially cancelled by the gravitational pull by Alpha Centauri. By then, though, the damage of this small velocity decrement is done, because the essentially terminal speed is incrementally lower than found in the absence of gravity in the models.

So, to continue the analysis, we must find the driver flux and stellar flux effects on the sail as functions of the distance from the sun. Neither direct flux ever falls to the ambient level for the given scenario except close to the other star. Consequently, both stellar fluxes continuously supply a measurable differential pressure for the entire journey, but that small differential does go to zero at some distance between the two sources for the attained velocity. If we wished to be thorough in the analysis, we would include the ambient flux of ~ $3 \times 10^{-8} w/m^2$ that is present everywhere along the trajectory. However, including this simply complicates an already complicated analysis and adds little to the dynamic analysis other than to supply what would be a further perturbation on the direct stellar flux from the sun and Alpha Centauri, which is also a perturbation on the direct microwave driver flux…a perturbation on a perturbation, which we will note but ignore.

Now, we will redo the Starwisp mission calculations to include the differential force from the ambient stellar fluxes and compare the results to the earlier calculations. We will estimate the magnitude of these effects, so that we can validate calculating the effects as if they are perturbations. What this means is that we can find the Starwisp velocity from the direct driving results and use the resultant velocity profiles to scale the magnitude of the second-order ambient effects, which then simply become additive and have little if any impact on the first-order velocity calculations…which are a result of Doppler and compression effects. This approach is typical when we are simply perturbing a first-order result with a small second-order effect.

For these perturbation calculations, we need to find the net flux for all points along the trajectory spanning from the sun to the barycenter of the Alpha Centauri system. We then need to find the net radiation pressure and subsequent net forces and then the net accelerations from these stellar fluxes and perform the same integrations as before to identify the impact of these differential forces on the velocity profile for Starwisp. Even though the solar radiation acceleration is miniscule compared to that of the driver source, at some range from the sun, Starwisp is under miniscule direct acceleration and is essentially coasting, with the Alpha Centauri radiation supplying a negative acceleration. Since flux is additive, the radiation pressures and accelerations are also additive.

To summarize, we can find the net stellar-flux force acting on the sail by adding the solar Doppler redshifted flux pushing the sail to the Alpha Centauri Doppler blueshifted flux that is retarding the sail's motion including the appropriate compression factors. We could likewise assume that the other forward and rearward ambient stellar fluxes are equal. However, since the ambient flux is orders of magnitude less than the direct solar and Alpha Centauri fluxes along most of the trajectory, these other ambient fluxes will be ignored. We have made a rough estimate that the flux for the Alpha Centauri system is about twice the solar flux. We know the solar flux at Earth, so we can scale the flux from the sun and the Alpha Centauri system via an inverse-range square law to any distance from the sun. Therefore, at any distance r from the sun, we can find the flux from the Alpha Centauri system algebraically. But, to find the net pressure on the sail, we will also need to include compression factors and not just the net flux alone.

However, to establish that the stellar fluxes are perturbations, we need to produce some first-order estimates of the sizes of the fluxes we are working with. The two direct stellar fluxes are balanced when $(1/r^2) = 2/(R - r)^2$, where the factor of 2 arises from the fact that the Alpha Centauri system has about twice the luminosity of the sun and where R is the nominal separation between the barycenter of the Alpha Centauri system and the sun, which is ~4.37lys and r is the distance from the sun. Solving this equation shows that the fluxes are equal in magnitude at ~ 2.25lys from the sun. We can find the magnitude of the solar flux at this distance r using the simple ratio of $r_e^2 \times 1353/r^2 \sim 6.7 \times 10^{-6} w/m^2$,

Appendix 1—Dynamic Radiation Pressure Models

where r_e is the distance of the Earth's orbit where the solar flux is 1353w/m². This can be compared to the driver radiation flux at the stellar flux balance distance, which is ~ 10^{-5}w/m². Therefore, the ambient stellar flux of ~ 3×10^{-8} w/m² is indeed a small…and ignorable…perturbation on the direct stellar fluxes of ~ 6.7×10^{-6}w/m² and on the driver flux of ~ 10^{-5} w/m². Consequently, at the distance of 2.25lyr from the Earth, the hybrid Starwisp is experiencing a direct static driver flux ~ 10^{-5}w/m², and both forward and rearward static stellar fluxes of ~ 6.7×10^{-6} w/m². Note that these are fluxes and not propulsion forces, which are yet to be calculated.

Even though the direct stellar fluxes are nearly the same as the microwave flux at these interstellar distances, we are interested in the potential of these various sources of radiation to produce further accelerations (or deceleration) on Starwisp. And we do find that the direct microwave acceleration and the direct stellar accelerations do produce minor changes in the attained velocity, such that all the accelerations from these sources are simply additive perturbations onto the attained velocity. Looking at the model for Starwisp, we find that the essentially terminal acceleration and terminal velocity occurs very near the sun…within what we broadly define as the solar system. While the velocity continues to increase slowly along the entire trajectory, it only adds an inconsequential speed to the Starwisp, which we estimate below.

However, meaningful driver acceleration has essentially ceased far before the static flux balance point, and the Starwisp is, for all practical purposes, coasting along most of the trajectory with a driver acceleration of ~ 3×10^{-6}m/sec² at the static balance distance. And as confirmation of the size of the perturbation, we find that the static direct stellar accelerations which canceled are on the order of ~ $\pm 10^{-8}$m/sec² at the balance distance. We can surmise that the direct stellar fluxes have a negligible effect on the Starwisp's dynamics until the mission approaches the Centauri system, which satisfies the criteria for these effects to be additive perturbations.

We can also find the range and velocity when the Doppler and compression factors cause these two stellar fluxes to be equal, though this is not truly necessary other than as a check on various assumptions. All things being equal for 100% reflectivity on both sides of the sail, the net stellar pressures as a function of speed and distance can be found by applying the Doppler and compression factors to each side of the previous simple model for apportioning the flux between the two sources. The Doppler and compression factor is $(1 - \beta \cos\varphi)^2/(1 - \beta^2)^{1/2}$, where φ is the angle between the direction of motion and the direction of the flux propagation. For the solar flux, $\varphi = 0$ deg but for the flux from the Alpha Centauri system, $\varphi = 180$ deg. Therefore, we must solve $(1 - \beta)^2/r^2 (1 - \beta^2)^{1/2} - 2(1 + \beta)^2/ (R - r)^2(1 - \beta^2)^{1/2} = 0$ for r as a function of β. What we are calculating is the point at which the dynamic differential or net stellar flux acceleration goes to zero.

Since the mid-trajectory speed of the Starwisp is $\sim 0.166c$, we can use this speed to find the point at which the dynamic stellar radiation pressures balance… though this is simply a separate exercise to confirm that the dynamic balance point still falls in the mid-region of the trajectory. Using the value of $\beta \sim 0.166$, we find the dynamic balance range to be $r \sim 1.47$lys from the sun rather than the static balance point range of 2.25lys. At 1.47 lys, the direct driver flux is $\sim 2 \times 10^{-5}$ w/m^2, which is $\sim 2x$ the flux found at the static balance range. Consequently, the direct driver acceleration is $\sim 6 \times 10^{-6}$ m/sec^2 or twice the static balance-point direct acceleration. Now, it only remains to calculate the cumulative effect of the differential stellar radiations pressures. These side analyses demonstrate, though, that we are dealing with small incremental perturbations when the stellar fluxes are factored into the Starwisp dynamics.

We find the accelerations and net velocity associated with the differential pressures by subtracting the direct driving effects from the models. We can do this because the differential stellar contributions to the total velocity profile are so miniscule that the direct driving force is defining the speed β used to calculate the differential stellar contributions. However, the direct stellar flux is on the same order of magnitude as the microwave driver flux at these same ranges. Therefore, while these fluxes are only supplying small perturbations on the overall mission velocity profile, these fluxes are comparable in their accelerations over the majority of the mission.

To recap, the two segments of the direct acceleration are characterized by a constant acceleration followed by an acceleration that varies as the inverse-square range, and the two driver acceleration models are $\{\beta [(1 - \beta^2)^{0.5}/(1 - \beta)^2]\} d\beta = K dr/c^2$ and $\{r_o^2 \beta[(1 - \beta^2)^{0.5}/(1 - \beta)^2]\} d\beta = K dr/c^2 r^2$, where K was the initial static acceleration of 1130m/sec^2 associated with the driver flux of 10^4w/m^2 driving the total system mass. The quantity c^2 arises from the conversion of the variable from v to β. The value for K was found for a total system mass (payload plus sail) of 20 grams and a driver "efficiency factor" of $\sim 34\%$ arising from optical and electrical considerations, which, unlike for Landis model, is just a number that has neither been calculated nor measured.

Appendix 1—Dynamic Radiation Pressure Models

For our hybrid optical Starwisp, we need to solve the above equation for β as a function of r and then use these values of β to find the differential stellar flux acceleration, treating the small accelerations as a perturbation. All in all, this has become a messy exercise in keeping track of all the assumptions and hybridization issues. Ultimately, the point is to wrap the ambient radiation pressure models in some numbers that are reasonable in sizing the expected stellar radiation accelerations.

The above mix-and-match issues are why parametric analyses should always start from scratch to ensure internal consistency in the models. In the current circumstances, we are looking for the consequences of un-modeled physics, which would be refined within a serious parametric model. In addition, the modeling here is addressing technique, since the scenario is a fabrication. As we saw in Chapter 5, the analysis of a particle accelerator indicated that accelerators are the only current technology in which these ambient factors play a role in defining the performance of modern technologies. Of course, cathode ray systems are also impacted as are electron microscopes. In fact, cathode ray acceleration limits were some of the early suggestions that new physics was involved with the acceleration of high speed objects. Therefore, while most such technologies have become obsolete, we now have a better understanding of their physics.

The ambient radiation models are transcendental and to find a function for the velocity as a function of the range, we need to develop a polynomial fit to the calculated baseline for velocity as a function of range and then use this in the stellar flux model to find the radiation pressure perturbation as a function of range…a lot of work for a little. In other words, since the ambient fluxes are simply perturbing the baseline acceleration, we need the attained baseline velocity profile to calculate the perturbation at any position along the trajectory of the radiation sail, including ranges close to the sun and target star system.

The first-order polynomial fit of β as a function of range used in the analysis has a small error at the mid-point of the acceleration profile, which we can tolerate given the very small magnitude of the stellar flux accelerations during the latter half of the initial acceleration phase out to r_o. By making the polynomial fit using more terms, we could reduce the error as much as required…and as stated earlier, a lot of work for a little.

To determine if the effort has any payoff, we can look at a single point, the transition range r_e, when the baseline supplies parameters $\beta \sim 0.121c$ and

$r_o \sim 6.8 \times 10^8$ km. If the scenario starts at Earth's orbit, r_e, we would find the solar flux from the ratio of the squares of the ranges times the solar flux at Earth's orbit. Consequently, at r_o the effective solar flux $\sim$ 15w/m². The acceleration on the sail at Earth's orbit is $\sim$ 153m/sec² or $\sim$ 15.6g! At r_o the static solar radiation acceleration has been reduced to $\sim$ 5m/sec². These very large solar flux accelerations are, of course, the result of Starwisp having a sail plus payload mass of only 20 grams.

The model for the dynamic direct net stellar radiation acceleration over the initial phase of acceleration out to r_o from Earth's orbit is $\{2\, r_e^2\, e_F\, P_e / M\, c^3\, (1 - \beta^2)^{0.5}\}\, \{((1 - \beta)^2/r^2) - 2\,(1 + \beta)^2/(R - r)^2\}$, where the negative sign within the second bracket pair indicates that a force component from Alpha Centauri is slowing the sail, β is positive, and P_e is the total solar flux intercepted at Earth's orbit by the 1km-square sail. The effective flux from the sun has been reduced from the static values by Doppler and compression effects, which for $\beta = 0.121$ and r_o, yields a dynamic factor of 0.78 for the solar flux, reducing the acceleration for this flux at r_o to 3.9m/sec² for the Starwisp sail.

We find that the flux from Alpha Centauri at r_o is $\sim 3.6 \times 10^{-8}$w/m², which results in a static deceleration for the Starwisp sail of $\sim -7 \times 10^{-7}$m/sec². The dynamic compression and Doppler increase this by $\sim$ 27% or to $\sim -8.9 \times 10^{-7}$m/sec², which is six orders of magnitude less than the direct dynamic solar acceleration at the same range. Consequently, we can ignore this deceleration component and only use the acceleration from the solar flux, since there is no measurable deceleration component from the Alpha Centauri flux within the solar system on a Starwisp system. But, of course, if we were to calculate this retarding force, the resulting reduction in velocity would be integrated along the whole trajectory, which would impose a slight penalty on the total trip duration. We can get a quantitative feel for that later in the analysis.

The solar flux causes the initial Starwisp acceleration to be slightly higher than from just considering the direct driver flux. Therefore, we need to solve the complete static problem first and then add the Doppler and compression factors to see what the new velocity profile is out to r_o. The dynamic model used to find β for constant driver acceleration plus solar acceleration is $\beta\, d\beta = \{K\,(1 - \beta)^2/c^2\,(1 - \beta^2)^{0.5}\}\, dr + \{2\, r_e^2\, e_F\, Pe\,(1 - \beta)^2/r^2 M\, c^3\,(1 - \beta^2)^{0.5}\}\, dr$, where $K = 1130$m/sec² and contains the factor e_F, M, and c. Solving this model for small-β, we can find an approximation for β as a function of range that is accurate enough for our purposes.

Appendix 1—Dynamic Radiation Pressure Models

For the direct-driver radiation, which is the first term in double brackets, $\beta \sim [K\,r/0.592\,c^2]^{1/2}$, and putting in $r = r_o$ yields $\beta = 0.121$. The initial static model used by Forward and everyone else is just $\beta \sim [K\,r/0.5\,c^2]^{1/2}$, which comes from the classical Newtonian relationship $v = [2\,K\,r]^{1/2}$, which comes from combining $v = K\,t$ and $r = 0.5\,K\,t^2$ and eliminating t. The classic static model results in finding β at r_o as being 0.132. Multiplying the expression by the dynamic factors changes the numerical coefficient from 0.5 to 0.592, which reduces the new initial-phase dynamic velocity to $\beta = 0.121$ at r_o.

The solar dynamic acceleration can be found by solving the second term in double brackets: $\beta\,d\beta = 2\,r_e^2\,e_F Pe\,(1-\beta)^2/r^2\,M\,c^3\,(1-\beta^2)^{0.5}\,dr$, where $P_e = 1.353\times10^9\,w$ and $M = 20$ grams. The resulting model for β in this case becomes $\beta\,[(1-\beta^2)^{0.5}/(1-\beta)^2]\,d\beta = 0.383\,dr/r^2$, where the range is from r_e to r_e+r_o. We can then find the value for β at r_o to see what the contribution of the solar irradiance is on the velocity profile for the range out to r_o. However, we note that the model for the direct-driver acceleration must be modified, such that at $r = r_e$, both initial velocities are zero.

Without the dynamic factors, we find that the solar flux produces a static flux-velocity profile of $v = [2\,K'\,r_e\,(1 - r_e/(r_e + r))]^{1/2}$, where K' is the initial solar-flux acceleration of $153\,m/sec^2$ at Earth's orbit and where K' includes the mass, c, and e_F. The static solar irradiance produces an incremental velocity at r_o of $\sim 1.94\times10^5\,m/sec$ velocity, which is $\beta_o \sim 7\times10^{-4}$. We already know that the dynamic factors only decrease the final velocity, so that the dynamic β_o is even smaller than for the static case. Consequently, we can either use the direct-drive velocity profile to model β or we can simply ignore this small contribution. As stated before, a lot of work for a little.

We now look briefly at the static model including gravity as a negative acceleration factor. The solar acceleration of gravity at Earth is quite small at $\sim 0.0059\,m/sec^2$. We can use the static model developed above for the change in solar insolence on the sail with distance away from Earth, replacing the solar acceleration with the gravitational attraction, which gives $v_o = -[2K''r_e(1-r_e/(r_e+r_o))]^{1/2}$, where K'' is now $-0.0059\,m/sec^2$. We can stop at this point, since this is $\sim 3.9\times10^{-6}$ the magnitude of the solar radiation acceleration component and is, therefore, irrelevant to this analysis. Interestingly, this is also on the same order as the Alpha Centauri flux deceleration factor within the solar system, and we ignore both sources of deceleration in this modeling effort. The reason for the gravity being a negligible issue in the current scenario is that, with the large sail and small total mass, the solar radiation pressure is a dominant factor in the analysis of the dynamics for a Starwisp-like radiation sail. For a more realistic sail and payload mass, gravity becomes a more crucial factor in defining the radiation-sail dynamics.

The only measurable impact of the solar flux is to cause the initial acceleration to be slightly higher than that from just the direct driver flux. In this case, we need to find if there is an impact on the early attained velocities and, therefore, an impact on the travel time for Starwisp to reach r_o. During the range over which the solar flux can supply a meaningful acceleration, β is so small that we can further consider the solar flux to be well represented by the static model. As β increases, the solar flux falls substantially, thereby further enhancing the impact of the dynamic factors on the solar acceleration. Combining the dynamic acceleration terms and treating the solar component as a perturbation and essentially ignoring the dynamic elements, we have that the dynamic velocity profile of Starwisp from r_e out to $r_e + r_o$ is given by $v_T = v_d + v_s = \{1909 (r - r_e)\}^{1/2} + \{4.6 \times 10^{10} (1 - r_e/r)\}^{1/2}$, where v_d is the dynamic velocity components for the direct driver flux and v_s is the static solar flux velocity component.

We can find the transit time from r_e to r_o using the model $dt(r) = dr/v(r)$ or $dt(v) = dv/a(v)$, where (v) or (r) identify the variables as functions of v or r, respectively. Since we have all the terms as functions of range, r, including $\beta = v/c$, we will find time as a function of range. Since the solar acceleration and velocity are so small, we can use a power series to separate $1/v$ into two terms, one static and the other dynamic: $1/v_d$ and $- v_s/v^2_d$ and $t_T \sim t_d + t_s$, where $dt_d = dr/v_d$ and $dt_s = - v_s \, dr/v^2_d$, where subscript (s) is for static or solar and subscript (d) is for dynamic. We have, then, that while the dynamic factor increases the transit time out to any range, the solar flux works to shorten that time slightly. Fortunately, the integrals are analytic. Carrying out the integration from r_e to r_o results in $t_d = 36,210$ sec and $t_s \sim -337$ sec. The solar flux supplies sufficient acceleration to measurably shorten the transit time from r_e to r_o. A more detailed analysis also shows that the dynamic solar flux acceleration only results in ~ 2% decrease in the solar component of the velocity at r_o, which would translate into ~ +7 sec added to - 337 sec.

The purpose of the analysis is to show that the various dynamic attributes can have an impact on the dynamics of objects experiencing protracted radiation pressures. These effects are subtle. But, when more typical values of

Appendix 1—Dynamic Radiation Pressure Models

β are used, the small-β dynamic terms can introduce measurable perturbations in the dynamics of the objects, such as spacecraft and satellites, but only over long mission times. With the above conclusion, we can calculate the velocity profile associated with the direct stellar flux influences. The dynamics are such that the influences of the small fluxes occur over a long period. Consequently, there may be more impact that might be suspected…or maybe not.

We will perform the analysis over the range of the mission starting at 0.1 lys away from the sun and terminate the analysis at 0.1 lys away from Alpha Centauri. The model is a modification of one already used, which fully written is
$$v\, dv = 2\, r_e^2\, e_F\, P_e\, /M\, c\, (1 - \beta^2)^{0.5} [((1 - \beta)^2/r^2) - 2\, [((1 + \beta)^2/(R - r)^2].$$
Because the velocity of Starwisp is nearly constant at the terminal speed over the cited range, we can use a single value for β of ~ 0.166 to find the Doppler and compression factors as fixed values and then find, to first order, v as a function of r. In other words, the stellar flux supplies a small perturbation on the essentially terminal value for β. The function being plotted is
$v - v_i = \Delta v \sim \{153\, r_e^2\, [(1/r_i) - (1/r) - 2/(R - r_i) + 2/(R - r)\}^{1/2}$, where the subscript i represents the initial conditions at 0.1lys from the sun, and v_i is the initial velocity at 0.1lys from the sun and is taken to be zero in the above model, since we are only looking at the mid-course impact of the direct stellar radiation on the mission velocity profile.

The resulting velocity perturbation profile from ambient radiation is shown in Fig. A1.1 for the Starwisp sail mass. Even at r = 0.1lys from the sun, the solar flux still supplies sufficient flux to supply an additional acceleration of $\sim 20,000$m/sec, which is $\beta \sim 0.00007$ and which is a small perturbation on the essentially steady state value of $\beta = 0.166$. Given that Alpha Centauri is twice a luminous as the sun, the deceleration at Alpha Centauri is sufficiently swift that the solar flux acceleration component from r = 0.1 lys is cancelled at r $\sim$ 4.0035 lys from the sun or ~ 0.37 lys from Alpha Centauri.

We can summarize the findings that the dynamic ambient radiation effects are inconsequential to a vehicle such as a hybrid Starwisp-like radiation sail. As with the modeling used to find the time for Starwisp to reach r_o, we can make a similar calculation that shows that the solar flux shortens the transit time to Alpha Centauri by some small inconsequential time compared to the total trip time. The dynamic factors reduce the terminal speed while the solar flux increases this speed by a small amount. What we have ignored is that the direct radiation driver is supplying an additional and comparable acceleration on the sail, so that the actual velocity perturbations from ambient radiation are only important when the vehicle is close to Alpha Centauri. We have simply ignored the additional perturbation that the driving source is supplying to the velocity profile.

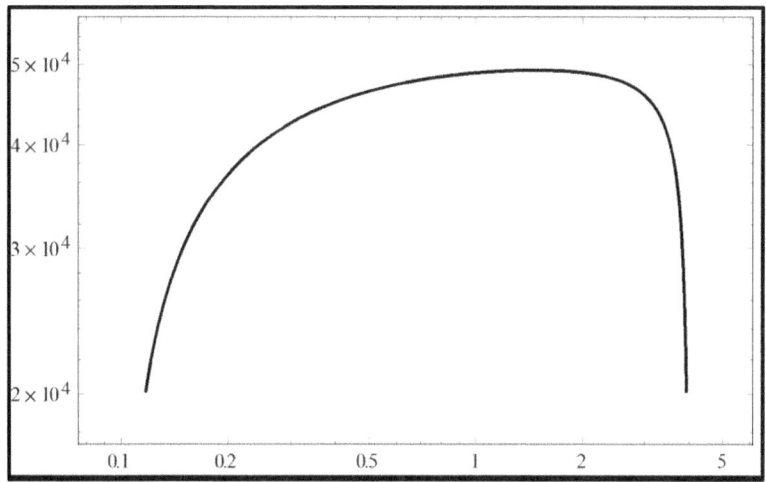

Figure A1.1 Stellar flux contribution to the Starwisp velocity profile from 0.1lys from the sun to 4.0035lys from the sun (~ 0.37lys from Alpha Centauri.

Figure A1.1 also indicates that a radiation sail, or any object, in interstellar space will eventually slow down absent some strong local source of radiation supplying a net positive or driving acceleration. A true interstellar or intergalactic environment will have a nearly uniform ambient flux in all directions. Hence, the models become simpler in that there are no range dependencies, only the velocity dependencies. The models are, therefore, like the initial acceleration phase models for Starwisp, scaled to the ambient flux levels. For speeds with $\beta < 0.1$, we can approximate the retarding forces with the model $(1 - \beta^2)^{.05} d\beta/\beta \sim - 8 A P_a dt/c^2 M$, where M is the mass of some perfectly reflecting object with a cross sectional area A and P_a is the ambient directionless radiation flux. This is simply the small-β description of the differential model when there are no inverse-range-square flux dependencies.

If we call $8 A P_a/c^2 M$ some constant K, then for small values of β we have that $d\beta/\beta \sim - K dt$, which when integrated yields $v \sim v_o \operatorname{Exp}(- K t)$, where v_o is the initial speed of an object at some arbitrary time zero. If we were to put in the parameters for Starwisp as it passed the Alpha Centauri system and assume it enters into interstellar space with an ambient omnidirectional flux of $P_a \sim 3\times10^{-8}$ w/m^2, then, since we know v_o, we can calculate K to be

Appendix 1—Dynamic Radiation Pressure Models

~ 1.33×10^{-16} m/sec². Of course, the hybrid Starwisp's value for β when passing through the Alpha Centauri system is ~ 0.166, which is not quite slow enough for the small-β approximation, but for estimation purposes, it suffices. And, of course, the true Starwisp does not have an ambient microwave flux to slow it down. Not surprisingly, the velocity decreases exponentially as with any viscous force retarding the motion of an object. For higher initial velocities, the retarding force is more complicated to calculate, since all the dynamic terms in terns of β must be retained. Still, once the speed has reached $\beta \sim 0.1$, the velocity decay is essentially exponential.

Since the velocity can never go to zero, there will always be some residual velocity characteristic of the ambient environment's kinetic temperature. Consequently, we can establish a scenario in which Starwisp enters some distance solar system with a residual local velocity of 10km/sec. Inputting a Starwisp initial velocity of $v_o = 0.1c$, we find that it takes ~ 5.7 billion years for the velocity to decay to 10km/sec via the differential Doppler pressure produced by ambient stellar radiation. In that time, the Starwisp has traveled ~ 70Mly.

On the other hand, a Starwisp vehicle within the galactic core might experience a wide variety of accelerating and decelerating forces as it caroms through the core in what would become a random walk until the vehicle reaches the outer areas of the core and percolates into intergalactic space. Once Starwisp enters intergalactic space, the irradiance would drop by many orders of magnitude, thereby extending the drifting scenarios by billions of years and millions of more light years.

If we compare the slowing of Starwisp with scattering from a molecule, we could substitute…to first order only…for the mass and sail area of Starwisp with equivalent atomic masses and optical cross sections. The mean optical (Rayleigh) cross section of an atom can be given as ~ 10^{-30}m² with a mass for hydrogen of ~1.7×10^{-27}kg. Putting these into the same model as for Starwisp and setting the initial velocity per atom to be the mean velocity for a medium-hot interstellar or intergalactic medium at $T \sim 10^5$k, we can find the time for the medium to "cool" to $T \sim 10^3$k based only on the differential radiation-pressure slowing of the atomic hydrogen. Admittedly, the ambient flux in intergalactic space would be lower than in interstellar space but likely not too much lower than within the galaxy far from the core…such as at Earth. Using that ambient flux, we find the cooling time to be ~ 0.55 billion years.

Clearly, the retarding forces from the ambient radiation are insignificant for most conceivable velocities…but not all, as discussed in Chapter 5 regarding particle accelerators. The practical retarding forces are, of course, based on a rough assumption of the ambient optical flux and would be location dependent. If we were to look at an object such as an asteroid with an extremely high mass-to-projected-area ratio, the ambient radiation pressure accelerations would be insignificant over the lifetime of the solar system, certainly as compared to direct solar irradiance effects that do have an observable impact on the motions of planetary objects. Dust, on the other hand, is directly driven by solar flux and could experience some measurable effects from extra-solar ambient radiation.

However, we ignored the other ambient radiation impinging upon accelerated particles from all directions. Consequently, not only is there a direct Doppler effect, there is an additional aberration effect that further increases the net retro-forces of the ambient flux. Now that the analysis in this appendix has been completed, what we have are dynamic models that can be applied to modeling the impact of radiation pressure on space objects within the solar system, including spacecraft and satellites. We also now have more accurate models to understanding the perturbations on the velocity and trajectories for long-duration missions.

Appendix 2—Notes on the Doppler and Aberration Models

We expand upon and supply more details here that support the narrative in Chapter 5. These details do not entail advanced mathematics. However, the details do entail some mathematics that is important for those who wish to pursue the physics and consequences of the aberration in more detail. The basic elements of the Doppler and aberration models were discussed in the first part of Chapter 6. Such details will only be re-introduced here to supply continuity and independence to the appendix. We also discuss some physics that were briefly alluded to in Chapters 5 and 6 that relate to how the Doppler and aberration models might be derived and validated using a different approach than used by Einstein. We also mix in some philosophical ruminations and commentary.

The purpose of the discussion in this appendix is to enumerate new sources of perturbation that are ignored, mostly because from a practical perspective they are irrelevant…or nearly irrelevant. However, for scientific work and for some new high-resolution defense system, both active and passive, these sources of perturbation may have an impact on the effectiveness of those systems. At a minimum, these perturbations can now be recognized for what they are, and if they are an issue, their presence can be deterministically accounted for if not eliminated as perturbations.

To begin, from a philosophical perspective, I have repeatedly commented on the fact that Einstein did not develop measurement models, which has had consequences in his manipulations to achieve various special relativistic models that we have shown are wrong. If we further make a point that waves do not propagate in a vacuum, we are left with showing that a flux of photons produces the same results as if we used the electromagnetic wave and Poynting vector approach. This is important in that we know how to model and measure radiation pressure, which we will show is a pathway toward validating Einstein's Doppler model for any velocity.

My contention is that the wave approach is, in part, a mathematical mnemonic that produces descriptions of the real underlying physics. There are no "waves" in empty space but there are, according to Maxwell's equation,

Appendix 2—Notes on the Doppler and Aberration Models

waves within materials or when there are currents present, which are created by the interaction of electrons and photons, though I have my doubts. In a material in which electrons are mobile, such as in a metal, a flux of photons creates a flux of electrons. The wave equations are used as a description for both types of flux. If there is no wave in empty space, there is no reason to believe that the wave equation is any more real for a flux of electrons in a metal. In either case, the wave equation must simply be a mnemonic.

In a broader sense, we may not be fully aware of what the results of our classical models are telling us. For instance, in quantum mechanics, it was Max Born who saw that the solutions to Schrödinger's equations were really amplitudes that could lead to probability distribution or density functions. That observation has proven to make quantum mechanics more understandable, though it may ultimately have had the opposite effect…which is not of any consequence in this book. With this in mind, let's see how Einstein's Doppler model derivation might have looked from the perspective of measuring a flux of photons.

To make the comparison of approaches, we re-examine the complex phase describing propagation, where the propagation exponent was identified as $-i(\omega t - \mathbf{k} \cdot \mathbf{x}) = 0$, which for propagation along the x-axis becomes $i(2\pi \nu t - 2\pi x/\lambda) = 0$. If we divide through by 2π and multiply by h, Planck's constant, we have $i(h \nu t - x h/\lambda) = 0$. But $h/\lambda = p$, the momentum of a photon, and $h\nu = E$, the energy of a photon. If we divide through by t and since $x/t = c$ for a photon, we have $pc - Et = 0$, which is a conservation equation, since pc equals photon energy per unit time and Et is energy per unit time, which is a flux. (Since these equations equal zero, for ease of unambiguously writing them, I occasionally switch the order of the terms.)

Consequently, we can easily broaden the description of the complex phase to represent a flux of photons through which a moving observer, represented by the primed quantities, can be moving, and the flux can be allowed to simply flow past a stationary point in the x-direction, which are the unprimed quantities. Therefore, Einstein's model for finding the Doppler and aberration can just as well be written in terms of the flux passing a point and as measured simultaneously by a stationary observer and a moving observer. Thus, we have $px - Et = p'x' - E't' = 0$ for an observer moving along the x-axis, the direction of flux propagation. The Lorentz transforms allow this to represent a type of conservation equation. Fortunately, this form is more related to an actual measurement model than is the complex phase definition for a propagating plane wave. The notion that a plane wave is a plane wave in all moving frames lacks, to me, physical significance, more so since there are no waves per se in a vacuum.

However, there is more to the above recognition of a flux. Einstein's approach was to make the propagation generalized by introducing direction cosines to allow a more generalized propagation. If we generalize to include motion across a flux propagating in the x-direction, we have the vector dot product $\mathbf{p} \cdot \mathbf{x} = p\, x\, \cos\varphi$ as describing the x-component of propagation along the flux direction. Written out in full we would have the complex phase as $p\, x\, \cos\varphi - E\, t = p'\, x'\, \cos\varphi' - E'\, t' = 0$. In keeping with the fundamental idea in relativity, any motion effects occur along the direction of motion and not orthogonal to this direction. Consequently, even though we could form something called an orthogonal Doppler, which would be given by $p\, x\, \sin\varphi - p'\, x'\, \sin\varphi' = 0$, which is the decomposition of the complex phase orthogonal to the direction of motion, it is not clear if this has any meaning. On the other hand, the above form does have relevance to the aberration, which could be thought of as a transverse Doppler effect.

The analog to this orthogonal Doppler was Einstein's transverse mass, whatever that was supposed to be, that had a functional relationship with velocity different from the longitudinal mass increase with velocity. It may be that the aberration is not completely described by Einstein's aberration equation, since the transverse relationship may represent another boundary condition that has been omitted in solving for the Doppler and aberration expressions. This will not be pursued any further. The only reason for bringing it up was that the form of the complex phase forces the modeling to be linear in the direction of motion, but the aberration supplies a transverse force in the orthogonal direction.

But from prior modeling, if an observer is moving, the radiation pressure also depends on the compression factor. In the pressure models, we relate the pressure acting over time as a force accelerating some mass. Therefore, it would seem as if we can identify another way of finding a Doppler relationship empirically by accelerating an electron to relativistic speeds and measuring the velocity this mass achieves after some period of time or after traveling some distance, since there is no longer a mass-increase phenomenon.

Still, we would require the compression factor to be present in a radiation pressure model. The compression conserves photons so that we

Appendix 2—Notes on the Doppler and Aberration Models

are comparing apples to apples. If the flux is monochromatic, then while the compression factor helps define the total collected photons during a measurement, the Doppler shift defines the new energy for each photon. As noted in Chapter 5, we measured total power, which is a combination of the Doppler effect plus photon-counting. Our apparatus allows us to separate the Doppler component from the total energy or intensity measurement component of what we measure. Therefore, in principle, the compression factor included in the radiation pressure model is unnecessary in the reformulated complex phase, which as stated earlier is basically only looking at the Doppler component. We do not need the compression factor when making a measurement to find the Doppler. Thus, Einstein's approach and a flux approach would appear to be equivalent.

The notion of a flux versus the notion of discrete photons has not been particularly well explored, and when it is explored, the results have required the most advanced models we have derived, even for simple experiments such as Young's double-slit apparatus. It is obvious from the radiation pressure models that we can view the radiation pressure as a consequence of the integration of all incident photons onto a surface in one second. If we look at this at a more granular level, then we have individual photons interacting with the atoms…or more specifically the electrons…within an object. At a still more granular level we need to understand how discrete photons interact with moving electrons and, in the case of photons, how a simultaneous torrent of photons in a flux impact discrete objects such as electrons within materials. (The latter condition is met by the radiometry of such interactions.) Such models and experiments have been performed. We discuss one of these experiments and models, the Compton scattering experiment and model, in Appendix 3. However, this famous experiment and model may not be what they appear, which is a discussion limited to Appendix 3.

Consider an optical system, in which radiation incident onto a surface produces a pressure, and the aberration occurs during the interaction of this radiation with the transverse component of motion of the moving surface, just as modeled for inverse-Compton scattering. The aberration continues to propagate through the system to the detection modules within a sensor. Therefore, when the incident angle is θ and aberration angle is θ' from the first surface the photon encounters, this deviated angle associated with the direct motion of a surface in a flux, continues through the system to the focal plane, being incrementally increased with each reflective optical element within the system.

In Chapter 6 we noted that there is an unexpected impact from the aberration when we are making measurements with a multi-surface sensor system. We can develop a first-order model using the small-β approximations, whereas for scientific work we may need to retain the complete complexity of the general Doppler and aberration models. Some of these scenarios will be qualitatively discussed in this appendix and in Appendix 3. For now, we will evaluate the impact of the small-β models.

In the small-β approximation, since the optical-path deviation angle is given by $\delta \sim \beta \sin\theta'$, the largest value of δ occurs for $\theta' = 90$ deg and has a maximum value β. For small-β, we have an approximation for both the total aberration and frequency-shift accumulations as a flux passed through a sensor system to the detector. We will describe how and why this is so, though the discussion has a variety of uncertainties associated with it.

The difficulty in arriving at a model for how a sensor's design impacts perturbations in both the Doppler and the aberration associated with a measured radiation source has to do with uncertainties in the physics that is involved. Cahill describes certain of these effects but leaves others unresolved. Cahill noted that radiation passing through a stationary gas appears to experience a phase shift associated with an actual flow of this gas resulting from the Earth's motion in the cosmos. From the perspective of the photons, a stationary gas is flowing at the hidden velocity. The phase shift, which is equivalent to a frequency shift, is similar to the Fizeau effect in which light passing longitudinally through a flowing stream of water is also phase or Doppler shifted according both to the speed of the water flow and to the index of refraction of the water...or liquid. In the next sentence, Cahill also states that such a phase shift is not observed when radiation passes through clear crystalline materials, such as through typical optics.

Cahill's assertions raise some unresolved issues. First, the Fizeau effect is not understood and is often attributed to something called Fresnel drag, which is variously described using kinetic relativistic effects, such as the addition of velocities. Since relativistic kinetic effects are arguably incorrect, the underlying Fresnel-drag phenomenon is not understood, nor is there an explanation why the same effect is not seen in crystalline materials. It would seem that, unless the Fizeau effect requires some level of non-linear optical effects within a material, the lack of an interaction of optical radiation within a transparent material requires quantum mechanical considerations. Yet crystalline materials have an index of refraction, which requires that the photons do interact with all of the electrons within a material. Consequently, there is no ready

Appendix 2—Notes on the Doppler and Aberration Models

explanation for the difference in the interactions unless is relates to non-linear optical effects, which seem to require specific types of polarizability of radiation within a solid material. This being the case, we can essentially ignore certain physics when we talk about radiation being detected and measured by a moving sensor system.

The hypothesis is that radiation entering an optical device can experience Doppler and aberration effects associated with the number of reflective surfaces that the radiation encounters in passing from the entrance aperture of the instruments until the radiation reaches the detector. In the following assessment, we will assume that only reflection from mirrors and absorption at the detector experiences any aberrations. Reflections from transparent optics have not been identified as experiencing an aberration effect, though this may simply be that such effects have never been investigated. Similarly, aberrations from frustrated total internal reflections or from dielectric coating may also never have been investigated for these physics. Consequently, the analysis below only supplies a roadmap in understanding how certain of our telescopes multiply the aberration and Doppler. In addition, these increases may not have any consequences, depending on the particular optical system under examination.

An experiment from 100 years ago by A. A. Belopolsky showed that we can locally cause a multiplicative factor to arise to enhance the magnitude of the Doppler. Here, we will perform a different thought experiment to show another multiplicative effect by simply looking at two parallel mirrors moving in the same direction with a flux of photons incident at an angle into the gap between the mirrors, such that the beam experiences multiple reflections as it propagates from one end of the parallel-mirror pair to the other. This model would be like a laser cavity (an etalon) moving at some velocity at some angle relative to the propagation direction of the photons. In a laser, in practice, we tilt the mirrors to eliminate "walk off" of the beam, and we adjust the longitudinal location of one mirror to adjust the output laser frequency for maximum beam integrity. We calibrate the system to ignore the physics. However, a laser in some arbitrarily moving platform such as a aircraft or satellite might experience variable effects as both the orientation of the laser in the platform and the velocity vector of that platform changes and introduces new errors that could only be calibrated out on the moving platform in real time. In other words, calibration is not a static process and must be a dynamic process to eliminate what may be deleterious perturbations in the system because of dynamic orientation changes. We will show what the magnitude of these perturbations can be in a simple model for an etalon.

Going back to the simple parallel mirror model, we can estimate the accumulated Doppler and aberration perturbations associated with the parallel mirrors using the simplistic model for a laser cavity. Envision parallel mirrors moving with some velocity, in which a beam of radiation entering the gap between the mirrors at some angle. The radiation would bounce back and forth and works its way down the gap and exit the mirrors at nominally the same angle relative to the direction of motion as the beam entered the gap in the first place. The Doppler shift would occur at the first surface and for parallel mirrors would have an equal and opposite Doppler shift at the second reflection. Depending on the number of bounces, the exiting beam will either have a single Doppler shift or no Doppler shift.

Now, if there is some aberration present, which there will usually be to one degree or another, the incident angle of the beam after the first reflection is at some new aberrated angle, and the Doppler shift at this second surface does not exactly counter that from the first mirror. Since the aberration accumulates one surface at a time additively, by the last reflection, there will always be some new Doppler frequency that is never cancelled. Again, depending on the number of reflections, the exiting beams will always have some incremental frequency shift that depends on the number of reflections that occur. If the motion of the mirrors is in the same general direction as the flux, that is, the incident beam enters the mirror pair from the backside, the aberration is in the direction of motion, so that the incident angle onto the second mirror is larger than the initial incident angle and the resultant Doppler is less than the initial Doppler. This accumulates until the beam exits. If the incident angle is from the front or opposite the direction of motion, the aberration is still in the direction of motion, but the aberration makes the incident angle onto the second mirror less than for the first mirror, which leads to an incremental increase in the frequency, which accumulates one surface at a time.

In general, unless an optical system contains an etalon, any multiple back-and-forth reflections are accidental, which might occur in a refractive optical system. For a reflective optical system, light simply propagates from the primary to the secondary mirrors, in which the Doppler and aberration from the first two mirrors are cumulative. Using Einstein's original Doppler model, we have that $\cos\varphi$ becomes $\cos(\varphi + n\delta)$, where $\delta = \beta \sin\varphi$ and n is the number of aberrating surfaces. Therefore, $\delta_n = n \delta_o = n \beta \sin\varphi$, where δ_o

Appendix 2—Notes on the Doppler and Aberration Models

is the aberration from the first surface. Putting $\cos(\varphi + n\,\delta_o)$ into the Doppler model, we have that the incremental frequency shift due to n aberration events is $\Delta v_a' = -v_o\, n\, \beta^2 \sin^2\varphi_o$.

This is a first-order idealized system model. The issue is that while a reflective telescope has two mirrors and the detector, it may also have a spectrometer, in which reflecting surface are not nominally parallel as with the two main mirrors. Thus, the angle of incidence relative to the direction of motion changes away from that for the idealized configuration. We would have to use a ray-tracing program to find the incident angle onto each reflective surface and from that find the aberration and incremental Doppler shift for each surface. We would then accumulate these changes one surface at a time to find the cumulative change for the entire system.

We can see that when or if the aberration is important in any given measurements, the total Doppler shift would also need to include the small second order term $-v\,\beta^2/2$. In this case, we would go back to the complete Doppler expression of $v' = \gamma v (1 - \beta \cos\varphi)$, where $\gamma = 1/\sqrt{1-\beta^2}$, and insert the total aberration shift before expanding the expression in the small-β approximation. This gives the complete Doppler shift as $\Delta v \sim -v\,(\cos\varphi_0 + n\,\beta^2 \sin^2\varphi_0 + \beta^2/2)$. The consequences are that there would be different frequency shifts for differently configured sensor systems.

We also identified a true zero Doppler angle when viewing the sun, which would be measurable in solar spectral measurements and for which there is no Doppler shift. However, this angle was found only using the single-surface aberration. In any given instrument, we would likely have to use the n-surface model to find the actual pointing angle for which the Doppler shift is zero. However, a measurement of the true-zero Doppler would be a definite minimum for any instrument and a strong indicator that the aberration model is correct, at least for small-β. It is possible that for pointing angles away from the minimum that the cumulative Doppler shift for the system could partially explain the spectral line red shifts as the solar image is spectrally scanned limb to limb.

As a final analysis, I considered the aberration as it might affect imaging in a space telescope, so I used scenarios based on the Hubble Space Telescope (HST), which is an extraordinary collection of technologies. No matter how the telescope is repositioned when pointing toward some stellar object, the image is always aberrated, which is true even for a flat surface. The only criterion for eliminating the aberration is to be moving toward or away from a source of radiation. If the image is dynamically centered on the central pixel in the detector, the telescope is actually pointing at the aberration angle relative to the incident rays from the radiating object. For Einstein's aberration, each reflecting surface adds to the aberration. Therefore, from the model for multiple aberration elements within an optical system, Einstein's aberration could be many multiples of the standard single-surface aberration. The question is, what difference does this make regarding how the HST tracks objects and records images?

As orbital telescopes are designed with larger apertures and mirrors, the fields of view of these telescopes become smaller. The consequence of these design changes means that the aberration per orbit can exceed the field of view of the telescope. The impact on tracking some object is that vignetting will begin to occur for certain viewing directions and the telescope tube will begin to block incident radiation as the image is dynamically tracked to keep the guides star's image near the center of the telescope's focal plane. Vignetting causes the total number of collected photons to be reduced. For very faint objects, such vignetting could reduce the number of objects that can be detected.

It is possible that aberration can impact the star trackers on deep space missions. The star trackers are used to align a spacecraft's trajectory and orientation or attitude. Such dynamic controls are necessary to keep the line of sight between the Earth and the moving spacecraft within the pointing parameters necessary for communications and antenna pointing plus proper vectoring of any trajectory-correction thruster burns. As communications move to higher frequency systems for supplying higher data rates, the pointing accuracy and stability becomes more critical. Star trackers typically have much less resolution than the HST star trackers, but two star trackers calibrated the same but pointing toward orthogonal tracking star sets could experience an angular mismatch between the star fields from each. This would occur because if the aberration of the star field images is different, and if this gross pointing error is not recognized or compensated, there could be some ambiguity as to the exact attitude and orientation of the space craft. The caveat is that these small aberrations may be below the resolution of the star trackers and, consequently, not be an issue.

In deep-space craft, the issue is that the value of β does not cycle per orbit and is not part of the initial system calibration and alignment. Most deep space objects undergo gravitational assist maneuvers and, consequently, can have values of β that are one or two times Earth's β in magnitude but in an entirely different direction than Earth's orbital β. This is then a permanent bias on any given star tracker's optical imagery, which would be different in magnitude

Appendix 2—Notes on the Doppler and Aberration Models

from tracker to tracker. These represent potential opportunities for alignment and attitude errors that may slowly drift over time as the orbital arc of the spacecraft changes.

Beyond these particular applications for optical systems and the few academically interesting applications, there are few other instances when the aberration might bias measurements. It is hard to imagine that a microwave system would experience significant aberration biases, though the frequency shifts rather than the angle biases may become an issue and should be investigated, especially for high resolution measurements. We already alluded to the fact that resonant devices may experience drifting as orientations change causing β to take on a range of cyclical values, which holds for both microwave sources and lasers.

With the possible scenarios discussed above for the aberration's impact on system reliability and stability, there is one last system that may or may not be a victim of unexplained aberration-based perturbations. We can investigate whether a terrestrial scenario exists for the impact of the aberration. Our calibration and alignment processes essentially negate all but diurnal and annual Earth rotation and orbital effects, and the rotation only supplies a variable value for β based on < 0.5km/sec rotation speeds.

On the other hand, we do have airborne platforms with laser system aboard which typically have variable values of β in the same speed range as the Earth's rotation, with speeds typically less than 1km/sec. However, the orbital location of the Earth would also impact the optical system, since alignment occurs in a particular location and with a particular device orientation. Therefore, we have similar aberration effects in both the pointing and tracking system and in laser beam optics, both internal to the laser and in the external optics. The scenario would fit that for the failed airborne lasers weapon platform.

It is not necessary to go into the details of the system, since there are simply too many sources of potential failure or perturbations due to the aberration effect. Not only would the physical pointing and tracking be affected, the laser itself would be subjected to angular and frequency-shift effects. All in all, the physics and technology of these systems was impressive but failed to achieve useful performance even after nearly forty years of effort. After the above analysis, it is possible that unknown physics may have contributed to the failure of these systems to perform.

The above discussions on the impact of the aberration points out that we compensate very well for any of the un-identified phenomenon, such as hidden velocities, Doppler effects, and the aberration. It is primarily when scientific measurements, most likely in a laboratory setting, are being made that the consequences of these phenomena are of importance.

To summarize, there may be, as noted above, certain modern technologies which are affected by the aberration effect, but these would be highly specialized systems with high-accuracy requirements that are inexplicably not operating steadily. These systems might be star trackers or various pointing and tracking systems, where operationally certain errors creep into the operation of these devices, causing deteriorating functioning of the systems. While these systems may operate on a rock-solid basis when bolted to the Earth and calibrated where they are tested, when we put these systems on moving platforms, the various "hidden" phenomena may cause the systems' performances to either degrade or fluctuate in unacceptable ways.

Appendix 3— Compton's Scattering Model

The relativistic discussion in Chapters 2 and 3, in which Minkowski's formulation of special relativity was heavily critiqued, raises the issue of where was Minkowski's approach used in other areas of physics. Specifically, the Minkowski form for the kinetic energy is not even part of the original relativity, but it was used to support three important areas of physics. One area is the relativistic formulation of quantum mechanics, another area is to support Compton scattering, and, the final area, is to support relativistic kinematics.

Compton scattering has been a bedrock of fundamental physics since the success of Compton's famous experiment in 1923 showing that photons were discrete entities similar in some ways to particles. Compton's model and experiment reinforced two aspects of Einstein scientific research. One was a validation of the corpuscular and quantized nature of photons, and the other was the seeming indispensability of the Minkowski kinetic energy in producing Compton's remarkably simple scattering model. While the first was clearly a viable accomplishment, the second assertion is not necessarily as firmly established as it may appear.

My inclusion of Compton Scattering is not simply to reinforce the idea that relativistic kinetic energy is not what we think, I included it because it is an experimental method of getting a handle on the Doppler expression developed by Einstein. I have alluded to the fact that the tests for the relativistic Doppler are nothing of the sort and validation is only for relatively slow objects. It would be easy to redo these low-velocity experiments at relativistic velocities, but the models used at these velocities would be heuristics and we have already highlighted the risks of using heuristics beyond their intended ranges of validity. So, while Compton scattering has been used with relativistic models, we already know that the kinematics and interactions are Doppler-driven using the Lorentz boost factor to include the form $1/\sqrt{1-\beta^2}$ that is necessary to introduce the velocity-limiting features into these models. In other words, these models have been developed under "false colors."

As has been noted, Minkowski's kinetic energy was essential in allowing Compton to develop his simple model, but that model may have only been

Appendix 3—Compton's Scattering Model

an apparent success. There is evidence that the choice of the experiment was instrumental in validating the model. The model does not appear to be generic and is, apparently, only approximately correct, though it is correct enough to explain the results of Compton's x-ray scattering experiment.

A quick summary of Compton's experiment is that Compton directed x-rays of a specific wavelength into a stationary block of graphite and, with a clever spectrometer arrangement, could measure the scattered photon as a function of the scattering angle relative to the direction of the input x-ray flux. The purpose of the experiment was to show that photons behaved as particles when scattered off atomic electrons, and the experiment and the model for it were based on a billiard-ball scattering model. The electrons in Compton's model are effectively stationary.

Compton developed a scattering model based on Minkowski's kinetic energy. The model is given as $\lambda_s - \lambda = h(1 - \cos\theta)/m_e c$, where λ_s is the wavelength of the scattered x-ray and λ is the wavelength of the incident x-ray, h is Planck's constant, θ is the scattering angle relative to the direction of the initial x-ray flux, c is the speed of light, and m_e is the mass of the electron that is scattering the x-ray. We can see that this is a direct analog to the billiard scattering models we talked about previously. The conservation of momentum and energy were used to develop the model, where Compton used the kinetic energy as given by Minkowski. The details of Compton's scattering are nicely laid out within Wikipedia. However, there are some caveats that come into play when we redo Compton's model excluding Minkowski's kinetic energy and simply use the classical kinetic energy.

The new classical Compton model development follows that usually described for developing the Compton model, but instead of introducing the relativistic features, the classical features were retained. An analysis of the results of the Compton model for his experiment indicated that we were likely not actually using the precision of the relativistic kinetic energy to enhance the precision of the model. For the electron speeds and "apparent" masses that were calculated from the scattering, the precision of a classical kinetic energy would be nearly indistinguishable from the precision of the relativistic form in Compton's experiment for the x-ray wavelength used by Compton.

Consequently, several small-parameter approximations were used in re-deriving the billiard model, analogous to using the small-β approximations in the Doppler and aberration modeling to reduce the complexity of the final forms for the resulting models. We also eliminated all relativistic factors, such as the Minkowski kinetic energy, and the mass-energy relationship, $m_o c^2$. It was initially thought that the mass-energy relationship was a replacement for some hidden velocity, but that proved not to be so. Inclusion of a hidden velocity vastly complicates the conservation equations and the resultant determination of the scattered photon's energy. For Compton's x-ray photon, though, we found that the recoil velocity of the electron was $v = 2 h/m_o \lambda \sim 0.04c$, which is ~ 40 times the value of any hidden velocity components. Consequently, we eliminated the need for the hidden velocity in the re-derived a classical first-order Compton-scattering model.

Another approximation used was that the defined Compton wavelength, $\lambda_c = h/m_o c = 0.0024\ nm$, was sufficiently smaller than Compton's x-ray wavelength of 0.07 nm that λ_c/λ could be used in a series expansion. In solving the conservation equations for the scattered photon, λ', we must solve a quadratic equation, from which we then find $\Delta\lambda = \lambda' - \lambda$ to compare with Compton's model. Additionally, the energy of a photon having Compton's wavelength is $m_o c^2$, which happens to be one-half of the energy for the threshold production of a stationary electron and positron pair. (However, striking an electron with a high-energy photon does not stimulate pair production, which is usually stated as requiring a "close" nucleus for momentum conservation. Nobody knows for sure if the photon must strike the nucleus, but for such high-energy photons, that seems to be a requirement. How can we not know the exact physics of pair production?

One interesting result of the billiard model is to supply negative wavelength solutions. In the simple billiard ball models for colliding masses, the negative values of velocity indicate a change of direction. In other words, even though we combine the momentum and energy models to find a final solution, the momentum is a vector and the solution to the quadratic equations preserves the vector directions of the solutions. Consequently, in a classical billiard model, we do not just find the magnitude of the final speeds, we also find their directions.

When we develop the new classical model for the scattering of a photon, we have a quadratic function to solve, which give Compton-like solutions. This is essential the same approach used to initiate development of Compton's own model. We use the momentum of the photons and the electron, including the electron energy as a momentum-squared term, staying strictly with momenta in setting up and solving the dynamic equations. Since the photon energy and momentum are related as $E = p\ c$,

Appendix 3—Compton's Scattering Model

the solutions to the model become substantially different from those for the kinematical billiard model. In fact, terms such m c and m c² appear because the speed of photons is a constant c, which replaces the more arbitrary velocities v in a billiard ball model. The terms m c and m c² are not particularly special and are simply algebraic expressions.

The general classical solution, following Compton's approach, is:

$$p' = p\cos\theta - m_o c \pm m_o c\sqrt{[1 - (p^2/m_o^2 c^2)\sin^2\theta - (2p/m_o c)(\cos\theta - 1)]},$$

where p' and p are the scattered and incident photon momenta, respectively, and the angle is the scattering angle relative to the incident direction of the photon. In the following analysis, we only use the back scattered photon, which is angle $\theta = 180°$, and the momentum of the back scattered photon is given as $p' = -p - m_o c \pm m_o c\sqrt{[1 + 4p/m_o c]}$. We see that we have negative solutions, which do not supply the complementary results to the Compton model and, so, will be ignored here, though they may be related to some other unrecognized physics. This is often typical in deriving models in which a quadratic must be solved. One of the sins of modern theoreticians is in forcing these unusual solution branches to represent new physics.

Substituting for the Compton wavelength and solving for the incident and back scattered wavelengths yields the following: $\lambda'^{-1} = -\lambda^{-1} - \lambda_c^{-1} \pm \lambda_c^{-1}\sqrt{1 + 4(\lambda_c/\lambda)}$, which is a general result for the back-scattered photons from a stationary electron. This is an algebraically difficult expression to render into $\Delta\lambda$, so we work with the momentum expression above. Changing the momentum expression to an energy expression using the fact that for photons $E = pc$ and using the result that $\Delta E = -hc\Delta\lambda/\lambda^2$, and, finally, expanding the resultant radical for $\lambda_c/\lambda \ll 1$ and keeping the first three terms, we reproduce, for the positive branch of the quadratic solution, Compton's result: $\Delta\lambda = 2h/m_e c = 2\lambda_c$. The fourth term in the expansion, however, shows that there is a deviation from Compton's result using the billiard model. Keeping the fourth and fifth terms, we find that $\Delta\lambda = \lambda' - \lambda \sim 2\lambda_c\left(1 - 2(\lambda_c/\lambda) + 5(\lambda_c/\lambda)^2\right)$. The model shows that the differential scattering wavelength is not constant but depends on the incident wavelength, which is more intuitively expected and contrasts with Compton's model, where the back scattered photon always is given by $2\lambda_c$.

For the wavelength used by Compton, the model shows that the differential wavelength using the billiard model differs from Compton's result by a factor 0.045, which is substantially larger than the deviation that could be attributed to replacing the relativistic kinetic energy by the classical kinetic energy. However, the deviation was still smaller than Compton's experimental error and would not have been noted. For longer wavelengths, the classical and relativistic kinetic energies show an even smaller difference, allowing the new classical billiard model to become even more precise. For these longer wavelengths, the billiard model approaches the Compton result, in which the wavelength difference is essentially constant for all wavelengths.

Contrasting the high-energy scattering with visible-photon scattering is a straight forward extrapolation from using the modified Classical Compton Model. Visible photons are approximately 5000 time less energetic than Compton's x-rays and the wavelength is 5000 times larger than Compton's wavelength, which suffices for the models to be a simple classical conservation models, such as the billiard ball model. In fact, a classical photon scattering model exists called Thompson scattering and, in the limit,, both Compton's and the new classical Compton model reduce to this classical model.

We arrived at a model $\Delta\lambda = \lambda' - \lambda \sim 2\lambda_c (1 - 2 (\lambda_c/\lambda)^2)$ for the change in wavelength at visible wavelengths. Consequently, the billiard model for the back scatter for visible radiation differs little from Compton's model. While Compton's model has no photon energy or wavelength scaling, the classical billiards model shows a small wavelength scaling, which in the visible is essentially unmeasurable. However, classical scattering at visible wavelengths are so unenergetic that bound electrons cannot be dislodged and, consequently, the scattering mass is essentially infinitely heavy and no electron rebound is possible. Hence, classical models show no wavelength change. However, scattering of visible light from free electrons would emulate the Compton scattering showing a small resulting wavelength shifts required from the conservation laws.

The billiard correction term in the above model show that the differential scattering correction is on the order of five parts per million in the visible compared to the magnitude of the Compton scattering itself and is $\sim 10^{-7}$ the magnitude of a visible wavelength. Consequently, this correction is unmeasurable. Even using a frequency deviation, the deviation is only a few hertz in the visible, which is unmeasurable. Only because photon fluxes have similar massive numbers of photons are we, though integration, able to measure radiation pressures with visible radiation.

The results are more dramatic for shorter wavelengths and higher energies than the Compton wavelength photon. For wavelengths shorter than

Appendix 3—Compton's Scattering Model

the Compton wavelength, the billiard model breaks down or at least gives unusual results. By this we mean that the backscattered wavelengths become negative, which was the branch of the previous solutions for lower energy photons that we ignored. The physics is that at the Compton wavelength, the forward scattering of the electron exceeds the speed of light. While we do not necessarily get pair production for these higher frequencies, the behavior of free electrons is not consistent with the simple billiard model. Pair production requires the presence of a nucleon to allow for momentum conservation, so that we can only get pure photon scattering with no pair production if we scatter from a beam of free electrons...there are always nuclei present for electrons in solids. Consequently, we truncate the modeling at this point, since the physics for photon scattering from electrons at frequencies above the Compton frequency requires different modeling efforts.

The modeling has been for a stationary electron being struck by a photon in which the photon is transferring the maximum amount of energy to the struck electron. A flux of photon of the same energy would, after the first few scattering events, develop an electron current with a distribution of velocities in the direction of the flux but also with an angular distribution that creates a diverging flux of particles. After the first scattering events, the electrons would gain sufficient energy to produce an energy and spatial distribution in the back scattered photons.

We have, therefore, several measurement scenarios that could be used to assess the form or the validity of Einstein's Doppler model for relativistic speeds and to validate the new formulation for Compton scattering. In arguments that were made earlier in the book, we concluded that the Doppler shift is independent of the scattering events used to measure the Doppler shift directly. This occurs because, as per Einstein, the scattering occurs with a Doppler-shifted photon, and the energy of the Doppler-shifted photon determines the nature of the scattering event. For electron motion in the direction of the flux, the resulting backscattered photons would have an energy distribution determined by the incident photon energy and not determined by the Compton wavelength. We could measure either the backscattered photon distribution or the energy distribution of the scattered electrons.

Therefore, in principle, we have several ways to determine the nature of the Doppler expression and the Compton scattering. That is, the correct interpretation of the Doppler shift depends on the photon and electron scattering physics defined by either the modified Compton scattering or the original Compton scattering. We can, in principle, determine which Compton scattering model is correct and which Doppler model is correct.

In determining the actual efficiency of our particle accelerators, we would need a measurement of the actual speeds of the electrons. Then, by varying the ambient temperature over a wide range of temperatures, we can determine how efficient the driving flux is in accelerating the electrons at relativistic speeds. This approach eliminates scattering cross sections, except for those scattering events that may be velocity related that emerge upon quantum modeling of scattering. However, we can still assume that Einstein's Doppler model holds for the quantum scattering events. The difference is the amount of energy that enters the recoil event

To summarize, an argument given previously indicated that Einstein's Doppler expression may be accurate for all velocities. The argument went like this: in the moving frame, the incident photons are already seen as Doppler shifted. Then, a scattering event or events occur based on this Doppler shifted frequency. What these upshifted frequencies may be determines the type of scattering event. Consequently, we can decouple the Doppler shift from the scattering and treat them as separate events, which they are. Therefore, there is no reason Einstein's Doppler expression does not hold for all velocities. The only difference in observed results of scattering is in the magnitude of the scattering and the exact physics of the scattering process as a function of the relative photon energy. We have already shown that the modified classical Compton scattering is different from the scattering predicted using Compton's original model. This difference would show up in the amount of upshift a photon experiences and the concomitant amount of energy exchange with some object, such as an electron.

Appendix 4—Planck's Dynamic Radiation Law

This appendix became necessary as the analysis of the ambient radiation effects began to show deviations away from the accepted discussion relating to blackbody radiation. At this point, I will state that I believe that the classical thermal radiation model, which is over a hundred years old, is a heuristic. Either that or it predicts counter-intuitive results for relativistic speeds.

The radiometric consequences of Planck's radiation law for blackbody radiation were extensively modeled during the 1950 to 1980-time because of the need for identifying infrared signatures and transmissivity of the atmosphere. Infrared imaging sensors were being introduced and there was considerable interest in maximizing their effectiveness while understanding how to concomitantly minimize signatures in combat scenarios. As it turns out, what was being studied was essentially static radiometry, and in this book, we need to understand dynamic radiometry just as we need to understand the dynamic radiation pressure. These are two aspects of the same physics.

Planck derived a law named after him that describes the emissivity of an object that has some absolute temperature T degrees Kelvin. The law describes the spectral content of the emitted radiation resulting from the temperature of the object, which is called blackbody radiation. Blackbody radiation has a spectrum that varies as a function of both the temperature of an object and the frequency of the emitted radiation.

If an object is traveling through a blackbody flux, the incident radiation from the forward direction is up shifted in frequency. If, on the other hand, a distance object is approaching a stationary object, from the Doppler expression we know that the measured radiation is also upshifted. In both instances, the detected power at each object depends both on the Doppler and on the compression factor, where the compression factor increases the flux density for approaching objects and decreases the flux for receding objects.

We measure intensities of the received radiation across a range of frequencies or wavelengths to determine if the incident spectrum represents

Appendix 4—Planck's Dynamic Radiation Law

blackbody radiation at a given temperature. Only in the case of small-β, the estimated color temperature of an emitter is essentially correct as defined by the Planck radiation law, which we show below.

However, arguments on how the radiation is affected by a Doppler shift are developed using relativistic effects, such as how measurement apertures or a sensor's solid viewing angles are affected by relativistic kinematic effects. So, for very high relative velocities, we have two issues. The issue is, if the kinematics of relativity are non-existent, what can we say about the change in the effective power that is measured from a blackbody when there is some relativistic net speed between the source and the receiver?

The current relativistic argument states that we can find the total incident ambient or thermal radiation power that is either upshifted or downshifted in frequency by using the relationship for the apparent Doppler temperature $T_d = T_a (1 + β)^{1/2}/(1 - β)^{1/2}$. From the two values, the forward upshifted and the rear red shifted radiation, we can calculate the differential power that is acting to slow a moving object. Once we have the apparent temperature shift from the Doppler, we find the power, and once we have the power, we must multiply it by the appropriate compression factor based on the direction of arrival of the radiation relative to an object's motion through the thermal radiation. And, in what follows, we ignore the fact that at truly high relativistic velocities, the actual measurement process changes as would our interpretation of what is being measured.

After performing the calculations described in Appendix 1, we find a net differential pressure on an object using the currently accepted form for the Doppler shifted thermal radiation. For small values of β, the differential pressure becomes $- 8 β σ T_a^4$, whereas for higher speeds, we need the complete expression, and the negative sign signifies a net slowing pressure from the thermal radiation. This is the accepted approach, though the compression factor is missing now.

Without going into the complete radiometry, the Planck radiation law is given as

$$I(v,T) = \frac{2hv^3/c^2}{e^{hv/k_BT} - 1},$$

where I is the spectral radiance at a given frequency v from an object at some absolute temperature T, h is Planck's constant, c is the speed of light, and k_B is called Boltzmann's constant. Various Wikipedia articles provide a good introduction to blackbody radiation and Planck's law.

The integral of *I* over all angles in a hemisphere and over all frequencies yields the total blackbody radiation power emitted from a unit surface at some temperature *T*. In real circumstances, it is a bit more complicated than that, but the integral yields what is called the Stefan-Boltzmann law, which is $P = \sigma T^4$, where *P* is the total radiated power, σ is called the Stefan-Boltzmann constant and *T* is the absolute temperature. For large distances from the source of radiation, we usually reduce the radiated power to a flux that is characteristic of the color temperature of the object, such as a star. Knowing the surface area that is radiating and the distance from the radiating object, we can find a flux at some location, which is what we either measure or identify as producing some effect, such as radiation pressure.

So far, the models are static. The dynamic models are described as reducing to the form $T_d = T_a (1 + \beta)^{1/2}/(1 - \beta)^{1/2}$. By this we mean that if we have a blackbody at some temperature T, if either we are moving toward this radiant object or the object is moving toward us, the measured flux will be changed from P to P' where $P' = \sigma T^4 ((1 + \beta)^{1/2}/(1 - \beta)^{1/2})^4$, where we have left the compression factor out but the Doppler is upshifted because the object is approaching the observer…or vice versa. For very large β, we can see that P' increases substantially, which as noted before is the standard description.

However, if we look at the model for *I* again, we can see how the spectral radiance would change under the same scenario. What we would be doing is looking at the radiation from a source using a spectrometer to measure the wavelength, and if we measure a series of such wavelengths, the intensity model should follow a blackbody with an apparent increased temperature. Again, the compression factor would further increase the intensity of the measured lines, but all lines would experience the same intensity changes, and the spectral distribution shape would remain the same, since all frequencies are Doppler shifted by the same amount.

If we substitute some Doppler shifted frequency for the given frequency, we can call the factor $f = (1 + \beta)^{1/2}/(1 - \beta)^{1/2}$, which is the Doppler upshift at some frequency ν and *f* is just a number. When we insert this into the model for *I* we find that

$$I'(\nu',T) = \frac{(2h\nu^3 f^3/c^2)e^{-f}}{e^{h\nu/k_B T} - e^{-f}}.$$

Since as β approach 1 for speeds close to the speed of light, *f* becomes very large, so e^{-f} goes to zero and the denominator goes to a constant value. When we look at ($f^3 e^{-f}$) as β approaches one, this product goes to zero. Since ν was an arbitrary choice, the radiation from the object would go to zero as the speed approaches the speed of light.

The only reasonable conclusion is that Planck's law is a heuristic that only holds for small-β...if at all for any consideration of the Doppler. That being the case, the question is how does the power emitted from a blackbody change with the Doppler shift? This is straight forward to answer by noting that flux is just photons per second and that the photons have some frequency. Since all photons are Doppler shifted the same, clearly $P' = P f (1 + \beta)$, where f is the linear Doppler shift for each photon and $(1 + \beta)$ is the compression factor for the scenario. We do not know enough to assign some pseudo-temperature to the spectral distribution of the Doppler shifted blackbody radiation flux.

However, as discussed below, the color temperature should not be Doppler shifted but the blackbody curve is Doppler shifted but not to some new pseudo-temperature, and the shape of the curve remains that for an object at the given temperature. Consequently, we must use Planck's radiation law with some discretion, and one must question the assignment of some pseudo-temperature to the Doppler shifted flux. However, the flux is simply Doppler shifted to a new flux in the dynamic radiation pressure models.

In the radiation pressure modeling, we identified the ambient radiations as associated with a flux, so that the Doppler shift of the power is the correct approach, avoiding the assignment of color temperatures to the emitting objects. Yet we only used the temperature of the environment to assign a flux, which we again directly Doppler shifted without any recourse to identifying a new pseudo-temperature associated with the upshifted radiation.

When we look at stellar objects, we use the discrete spectrum to identify any Doppler shifts to which to assign some relative velocities between the Earth and the object. I am not certain of the value of assigning some pseudo-temperature to stellar objects based on the spectrally measured Doppler shifts. Consequently, I am not certain where these Doppler pseudo-temperatures are of use.

However, in the spirit of completeness, I used the Stefan-Boltzmann radiation model to find the upshifted power, including using the compression factor, as measured by a moving observer and found a comparable pseudo-temperature of a blackbody emitting the same power. Moving objects...source and observer...add their own Doppler shifts and their own compression factors, but I only used a moving observer measuring the flux from the sun at the Earth's orbit. I modeled the consequences of an observer moving toward the sun at Earth's orbit at 0.1c. We know the flux at Earth from the sun is 1353watts/m² and we know that the sun is effectively modeled as a blackbody at 5778K. In the posited scenario, the Doppler and compression cause the measured power to be 1.22 times the stationary value or 1645 watts. From this the pseudo-temperature is found to be ~6073K in terms of power but not in terms of the spectral distribution.

However, since all frequencies are Doppler shifted the same, the effect is for the color temperature to not change, since the blackbody curve would still be matched to the solar blackbody, but the intensity would be as if the sun is closer than it actually is. The new pseudo-range to the sun is ~130M km or 0.9 times the Earth's orbit. If we were to use the current model for Doppler shifting the temperature, we would have the temperature shifted by 1.22, which would result in a radiated power associated with the new pseudo-temperature of 1.22^4 ~ 2.22 times the sun's blackbody temperature. The conclusion is that modern ideas of how to handle Doppler shifting of blackbody radiation are simply wrong on all accounts.

However, in the various analyses in this book we did not use the Doppler-shifted temperature formulation for the dynamic blackbody radiation. For those cases in which we used direct stellar radiation, we assumed we were dealing with fluxes that produced incident powers and, in Doppler shifting the power directly, we avoided assigning a color temperature to the radiation. This approach allowed the compression and Doppler shifting to define the subsequent incident flux due to the motion of an object through the flux. Since all frequencies in an incident flux are both Doppler shifted and compressed the same, we have essentially accurately accounted for the dynamic factors in the radiation pressure without needing to resort to some pseudo-Doppler shifted temperature.

References

References

arXiv.org—Archive maintained by Cornell University Library of mostly physics and mathematics pre-prints of papers, many of which are never refereed or published and are of mixed difficulty and quality.

URL=http://arxiv.org

Belopolsky, A. A., "On an Apparatus for the Laboratory Demonstration of the Doppler-Fizeau Principle," Astrophysical Journal, vol. 13, p.15 (1901)

URL=http://adsabs.harvard.edu/full/1901ApJ....13...15B

Brillouin, Leon, *Relativity Re-examined*, Academic Press 1970

Brillouin, Leon, *Science and Information Theory*, 2nd ed., Academic Press 1962

URL=https://en.wikiquote.org/wiki/L%C3%A9on_Brillouin

Brush, Stephen, "Why was relativity accepted?", Phys. Perspective 1 (1999) 184–214. (A PDF file can be downloaded.)

URL=https://scholar.google.com/scholar?q=stephan+brush+relativity&btnG=&hl=en&as_sdt=0%2C36.

Cahill, R.T., "The Michelson and Morley 1887 Experiment and the Discovery of Absolute Motion," Progr.Phys. 3 (2005) 25-29
{arXiv:physics/0508174 [physics.gen-ph]}

URL=http://arxiv.org/abs/physics/0508174v1

References

Cahill(2), R.T and Kitto, K, "Michelson-Morley Experiments Revisited and the Cosmic Background Radiation Preferred Frame", *Apeiron*, Vol. 10, No. 2, April 2003

URL=http://redshift.vif.com/JournalFiles/V10NO2PDF/V10N2CAH.pdf

Cerf, Roger, "Dismissing renewed attempts to deny Einstein the discovery of special relativity," American Journal of Physics, 74, 818 (2006)
Below is the now broken link to the web page where this article was originally available:

URL=https://scholar.google.com/scholar?q=Christian+Marchal+relativity&btnG&hl=en&as_sdt=0%2C36

Chandrasekhar, S., *Newton's Principia for the Common Reader*, Clarendon Press: Reprint edition (June 12, 2003).

"Criticism of the theory of relativity"

URL=https://en.wikipedia.org/wiki/Criticism_of_the_theory_of_relativity#cite_ref-wazeck_3-1

Collins, Harry, Andrew Bartlett, Andrew, and Reyes-Galindo, Luis, Perspectives on Science 25(4):pp. 411-438 (July 2017)

URL=https://www.researchgate.net/journal/1063-6145_Perspectives_on_Science

Darigol, Olivier, "On the Genesis of the Theory of Relativity," Séminaire Poincaré, 1, 1 (2005)

URL=https://books.google.com/books?hl=en&lr=&id=sBKNejgLW2UC&oi=fnd&pg=PA5&dq=olivier+darrigol+genesis&ots=TMcopum70M&sig=D5lGTMHzrgIpKn6UrZYiq3unDBE#v=onepage&q=olivier%20darrigol%20genesis&f=false

Dover Press—Affordable books on science and mathematics at many levels of difficulty, many of which were classics in their time. Searches on topics such as applied mathematics, analytical mechanics, and relativity will supply a lifetimes reading in these subjects.

Einstein, A., "On the Electrodynamics of Moving Bodies," ("Zur Elektrodynamik begetter Körper,") Annalen der Physik, 17:891-921

URL=http://einsteinpapers.press.princeton.edu/vol2-trans/154?ajax
URL=http://www.fourmilab.ch/etexts/einstein/specrel/www/
URL=https://archive.org/stream/principleofrelat00eins#page/n9/mode/2up
URL=https://en.wikisource.org/wiki/On_the_Electrodynamics_of_Moving_Bodies_%281920_edition%29

Einstein, A., "On the Electrodynamics of Moving Bodies", Chapter 3 in *The Principles of Relativity*, Dover Publications, 1963

Einstein, A., Relativity: *The Special and General Theory 1916* (20th Ed. 1920)

URL= http://www.ibiblio.org/ebooks/Einstein/Einstein_Relativity.pdf
(Public Domain in the United States)

Famaey, B. and Binney, J., "Modified Newtonian dynamics in the Milky Way", Monthly Notices of the Royal Astronomical Society

URL=https://mnras.oxfordjournals.org/content/363/2/603.full

Feynman, R., Leighton, R., and Sands, M., *The Feynman Lectures on Physics*, Addison Wesley 1963. On-line version available from CalTech:

URL=http://www.feynmanlectures.caltech.edu/

"Fictitious Forces"

URL=https://en.wikipedia.org/wiki/Fictitious_force

Finkbeiner, Ann, *The JASONS: The Secret History of Science's Postwar Elite*, Viking Press, New York 2006.

References

Forward, R.L. (1985) "Starwisp: an Ultralight Interstellar Probe," J. Spacecraft and Rockets, Vol. 22, p. 345-350

URL=http://path-2.narod.ru/design/base_e/starwisp.pdf

Frank, Adam and Gleiser, Marcelo, "A Crisis at the Edge of Physics," New York Times, Grey Matter, June 15, 2015

URL= http://www.nytimes.com/2015/06/07/opinion/a-crisis-at-the-edge-of-physics.html?smtyp=cur

Goldstein, H., Poole, C., and Safko, J., *Classical Mechanics*, 3rd ed., Addison Wesley 2000

Google Scholar—Indexed Search Engine for most science, mathematics, and technology topics, uses a familiar interface, and is often integrated with a universities own library catalog for rapid search and acquisition of electronic versions of scholarly works.

URL=https://scholar.google.com/

Gordin, Michael, *The Pseudo-Science Wars*, The University of Chicago Press, Chicago 2012

"History of the Lorentz Transforms"

URL=https://en.wikipedia.org/wiki/History_of_Lorentz_transformations

Holtzman, J. A.et al, "The performance and calibration of WFPC2 on the Hubble Space Telescope", Astronomical Society of the Pacific, Publications (ISSN 0004-6280), vol. 107, no. 708, p. 156-178

URL=http://adsabs.harvard.edu/full/1995PASP..107..156H7

Kaiser, David, "How Politics Shaped General Relativity," NY Times, Grey Matter, November 6, 2015

URL=http://www.nytimes.com/2015/11/08/opinion/how-politics-shaped-general-relativity.html?_r=0

Kennedy, Pagan, "How to Cultivate the Art of Serendipity," New York Times, Sunday Review: Opinion, January 2, 2016

URL=http://www.nytimes.com/2016/01/03/opinion/how-to-cultivate-the-art-of-serendipity.html

Kitchin, C. R., *Optical Astronomical Spectroscopy,* Inst of Physics Publishing 1995

Landis, Geoffrey, "Advanced Solar- and Laser-pushed Lightsail Concepts," Final Report, May 31, 1999 NASA Institute for Advanced Concepts, 1998 Phase I Advanced Aeronautical/Space Concept Studies

URL=http://www.niac.usra.edu/files/studies/final_report/4Landis.pdf
(Get it while you can, since this is on a legacy site that may disappear at any time.)

Logunov, A.A., *"Henri Poincaré and Relativity Theory"*

URL=http://arxiv.org/abs/physics/0408077

Lubin, Philip, "A Roadmap to Interstellar Flight," Report to NASA under the NASA Innovative Advanced Concepts (NAIC) program: arXiv:1604.01356 [astro-ph.EP]

URL=https://arxiv.org/abs/1604.01356v7

Mallove, Eugene and Matloff, Gregory, *The Starflight Handbook,* Wiley Science Editions, John Wiley & Sons, Inc. 1989

References

Mandelberg, H. I. and Witten, L., "Experimental Verification of the Relativistic Doppler Effect," Journal of the Optical Society of America, Vol. 52, Issue 5, (1962) pp. 529-535

URL=https://www.osapublishing.org/josa/abstract.cfm?uri=josa-52-5-529 (Only the abstract can be viewed, but it contains the salient information.)

Marchal, Christian, "Henri Poincaré: a decisive contribution to relativity"

URL=https://view.officeapps.live.com/op/view.aspx?src=http%3A%2F%2Fwww.annales.org%2Farchives%2Fx%2FRelativity.doc

URL=http://gsjournal.net/Science-Journals/Research%20Papers-Relativity%20Theory/Download/5620

"Michelson-Morley Experiment"

URL=https://en.wikipedia.org/wiki/Michelson%E2%80%93Morley_experiment

MOND (The MOND Page)

URL=http://www.astro.umd.edu/~ssm/mond/

Nayhm, Luther, *Newton's Gravity*, Tarkas Press, Amazon

URL=https://www.amazon.com/Newtons-Gravity-Hidden-Heuristic-Crippled/dp/1537634054/ref=sr_1_1?s=books&ie=UTF8&qid=1492437599&sr=1-1&keywords=Luther+Nayhm

Persson, Anders O., "The Coriolis Effect: Four centuries of conflict between common sense and mathematics, Part I: A history to 1885," History of Meteorology 2, 1 (2005)

URL=http://empslocal.ex.ac.uk/people/staff/gv219/classics.d/persson_on_coriolis05.pdf

Popkin, G., "What it would take to reach the Stars", Nature 542, 20–22 (02 February 2017)

URL=http://www.nature.com/news/what-it-would-take-to-reach-the-stars-1.21402?WT.mc_id=TWT_NatureNews

"Relativity Priority Dispute"— A brief survey of more professional criticisms can be found in the Wikipedia article.

URL=https://en.wikipedia.org/wiki/Relativity_priority_dispute#Attackers_and_defenders

Rothman, T., "Lost in Einstein's shadow," American Scientist, 94(2), 112.

URL= http://www.americanscientist.org/libraries/documents/200622102452_866.pdf

Schaum's Outlines,

URL= https://en.wikipedia.org/wiki/Schaum%27s_Outlines

Shattow, G. and Loeb, A., "Implications of recent measurements of the Milky Way rotation for the orbit of the Large Magellanic Cloud", Monthly Notices of the Royal Astronomical Society

URL=https://mnrasl.oxfordjournals.org/content/392/1/L21.full

Sirianni, M., et al. "The photometric performance and calibration of the Hubble Space Telescope Advanced Camera for Surveys," Publications of the Astronomical Society of the Pacific 117.836 (2005): 1049.

URL=http://iopscience.iop.org/article/10.1086/444553/pdf

Siegel, Nathan, "Sorry, But Lasers Won't Get You to Mars Anytime Soon," Forbes Magazine science article

URL=http://www.forbes.com/sites/startswithabang/2016/02/23/sorry-but-lasers-arent-taking-you-to-mars-anytime-soon/#c9d5f303bf4d

References

"Synthetic Aperture Radar"

URL=https://en.wikipedia.org/wiki/Synthetic_aperture_radar#cite_ref-16

Thompson, A. R., Moran, J. M., and Swenson, Jr., A. W., *Interferometry and Synthesis in Radio Astronomy*, 2nd ed., Wiley VHC 2004

Vulpetti, G., Johnson, L., Matloff, G., *Solar Sails*, Praxis Publishing LTD 2008

Weinberg, Steven, Excerpts from "The Revolution that Didn't happen," *New York Review of Books*, Vol XLV, Number 15 (1998)

URL=http://www.physics.utah.edu/~detar/phys4910/readings/fundamentals/weinberg.html#back3

Will, C. M., "Was Einstein Right? A Centenary Assessment"

URL=https://arxiv.org/pdf/1409.7871.pdf